11-19-96

D1462122

THE PLAY OF NATURE

 **The Indiana Series in the Philosophy of Technology**
Don Ihde, general editor

THE PLAY OF NATURE

Experimentation as
Performance

ROBERT P. CREASE

INDIANA UNIVERSITY PRESS

Bloomington and Indianapolis

The paper used in this publication meets the minimum requirements of American
National Standard for Information Sciences—Permanence of Paper for Printed
Library Materials, ANSI Z39.48-1984.

 ™

Manufactured in the United States of America

Library of Congress Cataloging-in-Publication Data

Crease, Robert P.
  The play of nature : experimentation as performance / Robert P.
Crease.
    p.   cm. — (Indiana series in the philosophy of technology)
  Includes bibliographical references and index.
  ISBN 0-253-31474-7 (alk. paper)
  1. Science—Philosophy.   2. Science—Experiments—Philosophy.
3. Performance. I. Title. II. Series.
Q175.C885—1993
507'.24—dc20   93-2735

1 2 3 4 5   97 96 95 94 93

CONTENTS

FOREWORD

Scientific experimentation, practiced since the origins of modern science, has often been mentioned but rarely examined in depth by philosophers of science. In part, this is because of the exaggerated preference for theory that has been the traditional hallmark of philosophy of science, especially in its once dominant positivist forms. However, at least since the Kuhnian revolution more than three decades ago, new perspectives on science as a human and cultural phenomenon have proliferated. Whereas realist/anti-realist controversies previously held center stage, first the newer forms of the sociology of science with strong programs in "social construction" and later many other approaches claimed attention. This series, with its focus on science's technologies, has an obvious interest in instrumentation and experiment, which in effect represent the technological embodiment of science.

The Play of Nature: Experiment as Performance, by Robert P. Crease, reconfigures the context and understanding of experiment. In the process it addresses indirectly much of the current debate about the role of *realism* and *social construction.* Crease's project has several distinctive aspects. First, not unlike previous books in this series, it is grounded in the tradition of the praxis philosophers who decentered science's exaggerated emphasis upon theory. The ideas of Edmund Husserl, John Dewey, and Martin Heidegger form the point of departure for Crease's reinterpretation of experimentation. Second, Crease continues the series' emphasis on understanding science through its actions, which in experiment are complex and social but also embedded in instruments and technological complexes. Third, Crease does not dismiss the realism of science: even if scientific results are "produced," they are nevertheless "real."

The Play of Nature takes as its motivation the analogy between theater, as production and performance, and scientific experimentation. I confess that when Bob Crease first proposed this approach I was skeptical. He admits that others responded with similar skepticism. Why? In part because we tend to think of the *products* of science and theatre as being very different. Yet, as Crease successfully demonstrates, the task of preparing an experiment, its rehearsal, and ultimately its performance strongly resemble the experience of theater. Crease probes deeply and with great subtlety into the theatrical analogy for experimentation. The result is a rich, suggestive, and original way of understanding the social and instrumental result of science. Crease's approach is, moreover, informed not only by his broad philosophical background but also by his intimate dealings with the history of contemporary science. He is the coauthor (with Charles Mann) of a widely read history of modern physics, *The Second Creation,* and is the official historian of the Brookhaven National Laboratory.

We are happy to include *The Play of Nature* in the Indiana Series in the Philosophy of Technology. It exemplifies the balanced, critical, and pioneering thrust that we have taken as our goal for the series.

DON IHDE

ACKNOWLEDGMENTS

My two principal debts in writing this book are to Patrick Heelan, who waited patiently while I stumbled about trying to understand many of his ideas, and to Don Ihde, at whose instigation I first embarked on this project and who waited patiently while I completed it.

Babette Babich, David Pettigrew, and Lester Embree each provided me with opportunities to work out early versions of these ideas. Dare Clubb, Charles Mann, and Zev Trachtenberg not only read drafts of the book, but ever since we were all classmates in college together, have provided me with the kind of personal companionship and intellectual community that once upon a time universities took it upon themselves to foster. Other manuscript readers with valuable contributions were Robert Scharff, Stephanie Stein and David Weissmann.

THE PLAY OF NATURE

INTRODUCTION
THE PROBLEM OF EXPERIMENTATION

Contemporary scientists are apt to express impatience with the efforts of philosophers to understand what they do, finding these efforts unproductive at best and disruptive at worst. Steven Weinberg, who won the 1979 Nobel Prize in physics, once remarked that scientists' discoveries about space and time, causality, ultimate particles, and the like, "do not so much confirm or refute the speculations of philosophers as show that philosophers were out of their jurisdiction in speculating about these phenomena." Weinberg is not alone in his skepticism. Murray Gell-Mann, the 1969 laureate in physics, has claimed that undue attention to physics "muddies the waters and obscures [the] principal task, which is to find a coherent structure that works. Moreover, a philosophical bias may easily cause [a physicist] to reject a good idea." And few readers of *Surely You're Joking, Mr. Feynman!*, by 1965 laureate Richard Feynman, are likely to forget the scene in which the author, sitting in on a philosophy seminar, is asked whether he thinks an electron is an "essential object". Feynman's attempt to come to grips with the request leads him to pose the preliminary question of whether a brick is an essential object—which elicits a different response from each person in the room. "And just as it should be in all stories about philosophers," he wrote, "it ended up in complete chaos. In all their previous discussions they hadn't even asked themselves whether such a simple object as a brick, much less an electron, is an 'essential object.'" Still harsher allegations have been lodged by other scientists, who have accused philosophers of undercutting the authority of science, contributing to its demise as a positive force in our society, and thereby helping to spread pseudoscience and irrationalism.[1]

Those who examine the practice of science professionally can sympathize with these complaints. Can any other branch of philosophy be as prone to jargon, uncritical assertions, and unreflective thinking as that which has called itself "philosophy of science?" The field seems to invite loose talk about weighty matters. Its literature rarely shows either a real grasp of the practice of science or a knowledge of how to structure a philosophical inquiry into it.

The reasons are partly historical. Around the turn of the century, in Vienna, London, and elsewhere, the beginnings of what would become mainstream philosophy of science began to develop around a particular set of logical and epistemological issues. These issues were simply assumed to

1

be exemplified in the actual acquisition of scientific knowledge, making a
real look at it unnecessary; indeed, several canonical figures in that development explicitly stated their lack of interest in the actual workings of
science in favor of its "logical analysis" or "logical reconstruction."[2] The
logical and epistemological issues rarely had any connection with the way
science was actually practiced, and philosophy of science began to cut itself
off from what this branch of philosophy was supposed to be a philosophy
of. But if the body of inquiry that arose in this way is not a true philosophy
of *science*, then it is deficient as *philosophy* as well. One consequence is that
working scientists who peruse mainstream writings in the "philosophy of
science"—generally, from here on I shall drop the quotation marks, assuming that the reader knows how skeptical I am of the mission of the bulk of
the writings that go under this heading—are likely to be bewildered by
the lack of anything resembling an account of what actually goes on in the
laboratory, and perhaps even annoyed by the arrogance of authors who,
without making any attempt to observe actual conduct, feel free to pro-
nounce on the meaning and methodological foundations of science. Even
practicing philosophers can be frustrated by the special terminologies and
methodologies of philosophers of science, who have granted science a special and privileged status with respect to other kinds of human activities
and who thus see no need in their work to engage in an extended dialogue
with the philosophical tradition.

The Neglect of Experiment

The most prominent symptom of the failure of the philosophy of science
to heed the practice of science is its omission of a comprehensive account
of experimental inquiry. Philosophers have never denied the importance
of experimentation and have frequently lavished it with praise as a cornerstone—even *the* cornerstone—of science. Consider Kant's paean, in the
preface to the *Critique of Pure Reason,* to the experimental efforts of Galileo,
Torricelli, and Stahl. With their work, he wrote, "a light broke upon all
students of nature," for they discovered the futility of haphazard observations and learned that one must compel nature to comply with questions
put to it. Reason must not approach nature "in the character of a pupil
who listens to everything that the teacher chooses to say, but of an ap-
pointed judge who compels the witnesses to answer questions which he
has himself formulated. . . . It is thus that the study of nature has entered
on the secure path of a science, after having for so many centuries been
nothing but a process of merely random groping."[3]

But Kant and others have taken it for granted that the way scientist-
judges elicit testimony from witnesses is unproblematic, epistemologically
speaking, and that the principal work of science lies in formulating ques-

tions and interpreting answers. It is as if science were akin to the erstwhile television game show "Concentration," in which contestants make educated guesses regarding what questions are written on a hidden face of certain video cubes; no sooner is a guess formulated than offstage technicians activate equipment to reveal the hidden question. Philosophers may concede that the information on the hidden face, data, may need to be interpreted and sometimes can be quite Delphic. Nevertheless, that information is thought to emerge fully formed and complete, and the process by which it does so, experimentation, is regarded as more or less automatic and machine governed. It seems devoid of philosophical interest.

The absence of philosophical discussion about experimentation is astonishing and profoundly revealing about the shortcomings of present-day philosophy of science. Even high-energy physics, the branch of science where the role of theory is most marked, is dominated by experimental activity; Department of Energy figures show that experimenters outnumber theorists by about two to one, and experimental programs receive about twenty times the funding of theoretical ones. Yet *The Encyclopedia of Philosophy* has a major heading for "Laws and Theories" but none for "Experiment." *Reason and Nature: The Meaning of Scientific Method*, by Morris R. Cohen, an American philosopher influential earlier in this century, contains a single, passing reference to experiment in this index, while *Testing Scientific Theories*, a volume in a respected series in the philosophy of science, has no such reference and is concerned primarily with the relation between evidence and theories. Ernest Nagel's influential book *The Structures of Science* mentions experimentation only briefly. In his view, "the distinctive aim of the scientific enterprise is to provide systematic and responsibly supported explanations"; all that experiments do, evidently, is provide part of the *explanandum*.[4] Nagel and other philosophers of science do not deny that experimentation is important to science. But they clearly do not regard it as posing interesting philosophical issues.

No doubt the authors of such works could attempt to justify the lacuna by saying that they are treating only the logical or essential structures of science, and that while experimentation plays a prominent role in actual scientific practice by producing the observations on which theories are based, such prominence does not necessarily translate into an essential structure worthy of or amenable to independent philosophical analysis. The foundation is among the least visible features of a building, while many outwardly prominent features may not appear on a blueprint. But philosophers know to be skeptical of serenity regarding one's grasp of essential structures, and appreciate the difficulties standing in the way of access to them. As Aristotle liked to say, one needs to know whether one is on the way from or to the first principles; one must be sure that one has entered into the matter properly. By what right does one make a declaration about the essential structures of a thing when one has not prefaced the declaration with an exhaustive study of *all* its structures?

Aristotle was profoundly aware of the dangers of becoming too quickly confident about what is essential in a field, hence his emphasis on the importance of beginning any inquiry in the *hoti*—in how a thing appears or shows itself—and only then moving on to establish its *ti estin* or being.[5] Because only a glance at scientific practice reveals experimentation to be one of its most prominent features, it would seem only natural for a philosophical inquiry into the nature of science to take it as a starting point. In fact, we would have every right to distrust one that did not.

Recognizing the problem, a few philosophers, sociologists, and historians recently have devoted attention to experimentation and its role in science; these include Alan Franklin, Peter Galison, David Gooding, Patrick Heelan, Ian Hacking, and others.[6] Most of these attempts, however, share two principal and related shortcomings. Their approaches are not *comprehensive,* and they are not couched in a *language* adequate to the task.

A philosophical inquiry ultimately aims to provide an account of the phenomenon in question as a whole rather than treat a limited set of its aspects. This distinguishes philosophical accounts from sociological, historical, and economic accounts, for example, which attempt to approach their subjects within a single perspective. A philosophical perspective aims at comprehensiveness, or an ability to see things from a multitude of perspectives, and a philosophical grasp of the subject is lacking until we do so. When we scholars of the humanities trot out our awareness that science is a cultural practice among other cultural practices, and thus subject to economic, political, psychological, and social influences, and when we document that insight in well-researched papers, we may compliment ourselves on our lack of naïveté and on having gone beyond the traditional approach to science—but we have not yet achieved philosophical knowledge. It is one thing to discuss aspects of experimentation and another to discuss experimentation itself. It is one thing to conduct social, political, rhetorical, or economic studies into the planning, conduct, reporting, and reception of actual experiments; it is another to embark upon a comprehensive philosophical discussion about the nature of experimentation in general. To carry out the latter requires that we draw connections with at least some traditional philosophical discussions and use at least some existing philosophical tools.

I claim that a comprehensive approach to experimentation can be built with the aid of an analogy, what I shall call the *theatrical analogy.* Scientific experiments and theatrical performances have deep, striking similarities; moreover, the language of theatrical performance is comparatively well developed and solid enough to support the connections that need to be drawn. It is true that a certain feat of abstraction is required to see experimentation as a kind of performance, but no greater than the feats of abstraction that are an ordinary and accepted part of the practice of science. The analogy breaks down in places—the dissimilarities are as deep as the similarities—but with suitable adaptation can help forge (it is not a question of substituting for) a rigorous and comprehensive account of experimentation that does justice to the phenomenon. The use of this particular analogy is risky given the low esteem that philosophers have accorded theatre ever since Plato banished actors and other imitators from his ideal republic, and part of my task will be to generate an appreciation for the value of theatre and performance as worthy philosophical subjects. But the basic technique of using analogy as an argumentative tool is eminently sound. Analogies prove nothing, Freud (who used them often) observed, but they can make us feel more at home. They permit unfamiliar phenomena to be viewed in new ways by drawing connections, sometimes surprising ones, with things that are familiar; analogies are imagination-liberating, wrote Leibniz. Human beings, remarked Italian philosopher of science Giambattista Vico, understand the unfamiliar by the familiar, and Jeremy Bernstein opens one of his books with the remark that "it is probably no exaggeration to say that all of theoretical physics proceeds by analogy."[7]

The second principal shortcoming of much recent literature on experimentation is lack of a suitable language. The traditional language of the philosophy of science acts as a filter, facilitating discussion of certain aspects of science, chiefly those having to do with theory, while hampering and virtually obliterating others, especially those pertaining to experimentation.

With such a language, philosophical inquiry into science is doomed to fail. Trying to construct a philosophy of science with a language inadequate for addressing experimentation is like handing physicists hammers and nails and expecting them to do experiments in quantum mechanics or relativity. Even the most well-intentioned attempt to accord experimentation its due is doomed if it attempts to respect the constraints imposed by the traditional language of philosophy of science. And if one fails to understand the nature and role of experimentation, one is doomed to misunderstand science itself.

Consider a materials scientist working at an experiment in a lab. The scientist is tending to a complicated-looking device mostly covered with aluminum foil, arranging the position of an item in the device. Adjusting a series of dials and consulting a computer screen, the scientist modifies certain properties in the apparatus. More adjustments follow, and more modifications. What is happening, from the viewpoint of the traditional philosophy of science? Hardly anything. No data are beng collected, no hypotheses tested. There have been no verifications or confirmations. Eventually, measurements will be taken, data recorded in neat tables, talks given at other laboratories, a letter submitted to journal. But all that is weeks or months in the future. Shall we say that what is happening is preparation for the eventual "real" work of taking data? Nothing could be further from the truth. Already the theatrical analogy is helpful in providing a language; an action is underway, it is of the nature of a *performance* not yet perfectly executed, and the scientist is learning how to shape the performance of the instrument. Not much abstraction is needed to compare that scientist to the producer-director of a theatrical production preparing for opening night, after which performance after performance may be replayed with confidence and the results savored, evaluated, applauded, criticized, publicized, and appropriated in performances and productions of other kinds. Scientific inquiry progresses through the acquisition of the skills and instruments needed to produce such performances, and through their evaluation, standardization, and dissemination. But have we an adequate language with which to speak about them? Philosophers who cannot speak meaningfully about this kind of activity—who relegate it to the realm of propaedeutic to "real" science—deceive themselves if they think that they are speaking about science, and thus if they think that they are doing philosophy *of* science.

In this book, I claim that the theatrical analogy allows one to develop not only a comprehensive approach to experimentation but also an adequate language in which to speak of it philosophically. In chapter 1, I discuss the mythic approach to experimentation, and I raise some issues that need to be addressed in any philosophical approach to it. In chapter 2, I review existing philosophical tools that I find necessary in addressing these problems, even though they require adaptation. In chapter 3, I discuss the philosophical use of analogy, present a philosophical framework to serve as the background to development of the question of experimentation, and then introduce the theatrical analogy proper. Chapters 4 through 6 elaborate a philosophical account of experimentation with the aid of the theatrical analogy, each chapter devoted to one of the three essential dimensions of performance. Chapter 7 discusses production, a key concept for completing discussion of the issues raised in chapter 1.

Value of the Inquiry for Science

Unsurprisingly, scientists find that they rarely need to think about philosophical issues in the course of their professional work. Philosophical issues are almost never encountered in day-to-day activity or journal articles and rarely crop up even in memoirs or other popular works by scientists. The practice of science continues unabated in the absence of philosophical reflection, just as individuals can debate some topic without reflecting on the nature of the language allowing them to do so. When scientists do reflect on the nature of experimentation, it usually occurs when they depart from their professional activities to address lay audiences. The nature of experimentation thus does not seem to be an essential scientific issue; experimentation is something scientists have proven themselves able to do successfully and consistently without philosophical input. To put it in other terms: although scientists may face numerous *experimental problems*—finding, say, better ways to accelerate particles, more efficient light collectors, and more genetic markers—they do not appear to face a *problem of experimentation*.

What, then, could be the value of a philosophical account of experimentation? Scientists might find that it only "muddies the waters," in Gell-Mann's words. The average scientist, claimed chemist Gilbert N. Lewis in 1926, is a "nearsighted creature" who is "unequipped with the powerful lenses of philosophy." If philosophy made scientists self-conscious, they "might lose [their] power, like the famous centipede who, after too profound analysis of his own method of locomotion, found he could no longer walk."[8]

Such fears stem from a limited understanding of the nature and uses of philosophy. Scientific research takes place in a social and cultural environment and is therefore affected by the way it is perceived by the public; this perception affects, for instance, what kinds of science are funded and by how much. Lack of an adequate philosophical understanding of science has promoted perceptions of science that ultimately devalue and damage it. Positivists, whom I discuss in the next chapter, promote an image of science that science could never fulfill, leading inevitably to disillusion. This false image includes the assumption that the method of science is automatic and its knowledge independent of social and psychological forces, and that the picture of nature emerging from the laboratory is drawn in all its details by nature itself. Even newspaper accounts of scientific activity show a discrepancy between this picture and the actual practice of science, which can suggest that scientists are bunglers and frauds.

Let me give an example. In 1925, Dayton C. Miller attempted to repeat the famous Michelson-Morley experiment showing the constancy of the velocity of light regardless of the velocity of the source, a result that was the foundation for Einstein's special theory of relativity. Miller obtained contrary results, finding a slight difference in speed, and he interpreted the results as refuting Einstein. Reporting this to the American Physical Society, he was greeted with skepticism. "[T]he audience," claimed *New York Times* reporters William Broad and Nicholas Wade, writing years later, "should instantly have abandoned the theory or at least assigned it to provisional status."[9] The fact that they didn't, imply the journalists, means that the scientific community was guilty of incompetence and unprofessional conduct. But thousands of experiments had been performed in the forty years since the Michelson-Morley experiment with results in line with theirs, and Einstein's theory by then was becoming tightly woven within the fabric of contemporary science. Moreover, experiments are sometimes poorly executed, even by skilled researchers; as Martin Gardner remarks with typical wit, "it is always possible to find *someone* unable to perform an experiment."[10] To have suspended such a fundamental achievement as Einstein's because of a single contrary experiment would have been downright irrational. Broad and Wade's judgment of the delinquency of the scientific community is the product of an unreasonable but widespread portrait of scientists as akin to high priests, possessed of exceptional methods and insulated from the kinds of temptations and impediments faced by the rest of us. (It is not hard to understand why scientists themselves have been slow to disturb this picture.) Witness the sentiment of outrage and betrayal expressed in the title of Broad and Wade's book, *Betrayers of the Truth: Fraud and Deceit in the Halls of Science.* The disillusion that can accompany revelations of the falsity of scientific infallibility is manifested in the increasing desire by nonscientists to supervise research, especially medical research.[11]

More prevalent today, however, are views that devalue science by insisting that the study of the scientific process reduces to a kind of sociology or anthropology—to an analysis of the historical, cultural, and social forces that shape its origins and development.

Two centuries ago, Adam Smith was among the first to hint at a "social critique" of science by remarking that it could only develop when social "order and security" increased the curiosity and leisure time of human beings, making them "more attentive to the appearances of nature, more observant of her smallest irregularities, and more desirous to know what is the chain which links them all together."[12] More sophisticated analyses of the impact of social conditions on science in this century include the Russian Marxist Boris Hessen's analysis of the correspondence between the scientific problems Newton studied and the technical problems of the capitalist economy in which he worked, and Paul Forman's claim that European social currents of the 1920s shaped the development of quantum mechanics and its indeterminacy principles.[13] This general approach became hugely popular after the publication in 1962 of Thomas Kuhn's book *The Structure of Scientific Revolutions.* Indeed, *social constructivism,* as it is now known, is the dominant approach to the sociology of science today. Scientific knowledge is "underdetermined," say the constructivists; nature itself is insufficient to shape the picture of nature as it emerges in the laboratory, and the social behavior of scientists is the principal determinant of their picture of nature. Understanding science thus entails, first of all, understanding the social forces at work on scientists.[14] To cite one example, Margaret Jacob, in her study, *The Cultural Meaning of the Scientific Revolution,* points out the importance of English revolutionaries in creating an intellectual climate in which serious opposition to Cartesianism (then strongly held throughout continental Europe) could be maintained. "In this sense we may say," she writes, "that while the culmination of the Scientific Revolution is unthinkable without Newton, Newton is unthinkable without the English Revolution."[15]

The huge divergence between the positivist and the social constructivist views of science is emblematic of what might be called the *antinomic* character of scientific knowledge. Scientific knowledge is on the one hand a social product, the outcome of a concrete historical, cultural, and social context, and is therefore affected by the presence of political, economic, institutional, psychological, and other forces that are always at work in such contexts; on the other hand, scientific knowledge has a certain objectivity or independence from these contexts. (As we shall see in chapter 4, this antinomic character ultimately stems from the antinomic character of experiments, which are at once utterly anomalous occurrences executed within a highly specialized material context and at the same time manifestations of phenomena that transcend particular contexts.) Contemporary philosophy of science, whether social constructivist or positivist inspired, lacks the resources to describe this antinomic character within a coherent framework, and as a result, contemporary cultural discussions about science have been afflicted by a profound confusion about its nature. Social constructivism, for instance, has been interpreted as denying the validity of science.

This is far from an academic problem, as is suggested by a controversy about it that erupted in the pages of *Nature* in 1987. Two physicists from the Imperial College of Science and Technology in London claimed in an article that the "most fundamental, and yet the least recognized" cause of the dire predicament of British science was an "epistemological relativism"—their word for social constructivism—that had become fashionable since the publication of *The Structure of Scientific Revolutions.* Proponents of such relativism, said the authors, emphasize the transience of scientific theories and the importance of nonscientific factors in the acceptance and rejection of theories and in the conduct of science in general, and assert that science can claim no particular superiority for its method or results with respect to other cultural practices. The authors referred to such "erroneous and harmful ideas" as "epistemological antitheses," or "(un)philosophical positions which are contrary to the traditional and successful theses of natural philosophy," and asserted that their popularity had hurt the position of science with respect to other disciplines. "Having lost their monopoly in the production of knowledge," the authors stated, "scientists have also lost their privileged status in society. Thus the rewards to the creators of science's now ephemeral and disposable theories are currently being reduced to accord with their downgraded and devalued work, and with science's diminished ambitions." Quoting Kuhn's famous remark that "we may have to relinquish the notion that changes of paradigm carry scientists . . . closer and closer to the truth," the authors wrote that "by denying truth and reality science is reduced to a pointless, if entertaining, game; a meaningless, if exacting, exercise; and a destinationless, if enjoyable, journey." They concluded, "It is the duty of those who want to save science, both intellectually and financially, to refute the antitheses, and to reassure their paymasters and the public that the genuine theories of science are of permanent value, and that scientific research does have a concrete and positive and useful aim—to discover the truth and establish reality."[16]

Yet providing such reassurances is more difficult than the authors make it sound. The work of social constructivists, epistemological relativists, and others who pursue social studies of science is quite solidly grounded in empirical research. What does it mean to say that genuine theories are of "permanent value," and that scientific research discovers "the truth" and establishes "reality," in the face of the simple historical fact that theories as well as experimental methods, techniques, and equipment are constantly changing, and that science is clearly affected by the resources that a society chooses to devote to it? From the newspaper accounts that appear every so often announcing that some new measurement has doubled or halved calculations of the accepted age or size of the universe, or that some new discovery has made scientists abruptly change their minds about whether dinosaurs were warm- or cold-blooded or "somewhere in the middle" (not to mention what made them extinct), one can easily gather an impression of the transience of scientific truths.[17] Consider how radically our notion of the "fundamental unit of matter" has changed in our century alone, from uncuttable atoms to protons and electrons to quarks and leptons; scientists cannot guarantee that it will not change equally radically in the next. Martha Graham, the great modern dance innovator, surely can be pardoned for having remarked that "even the most brilliant scientific discoveries will in time change and perhaps grow obsolete, as new scientific manifestations emerge. But Art is eternal."[18]

By asserting that scientific truth is *not* transient and that it *does* deliver truths that are somehow independent of specific cultural and historical contexts, many scientists reject the claim of social constructivists, epistemological relativists, and other scholars of the humanities and social sciences who describe science as being shaped by social and historical factors. Nonetheless, it is impossible to doubt the legitimacy of studying how the growth and development of science has been affected by forces that are political (government's wartime interest in antibiotics, explosives, computers, codes, etc.), economic (the profit motive of pharmaceutical, energy, communications, and other industries), humanitarian (the desire to cure AIDS and cancer), social (the existence of practices such as beer fermentation techniques, which were easily transferable to producing penicillin), psychological (individual motivations for pursuing research), and of many other kinds.

What remains to be described, however, is the meaning of these studies. Through science, humanity manages to discover, stabilize, and produce new phenomena (antibiotics, microwaves, lasers) in such a way that they can be reproduced in widely different contexts and even with such regularity that the entire process can be taken over by nonscientists. If scientists wish to refute the disparagements of the Martha Grahams as well as those of the social constructivists, they will have to show clearly and persuasively how it is possible that science can be just as much a cultural product as art, politics, and philosophy, and how it can also manage, in the words of the *Nature* article, to "discover truth and establish reality." But the resources for doing so are unavailable in the traditional account of science.

Today, the mill of social studies of science scholarship grinds very fine. I am not a worker in that mill. I am attempting to do philosophy rather than sociology of science—which means, in this case, attempting to develop the widest possible perspective on experimentation. I aim to see the links, as it were, between the social studies of science mill and other mills, as well as the links between it and the production of grain, the uses to which the flour is put, and so forth. I aim, that is, to place social studies of science among various other kinds of studies that can be undertaken on the activity of science as part of a philosophical inquiry on experimentation.

To paraphrase a remark Marx made about society in the *Eighteenth Brumaire of Louis Bonaparte*, human beings make science, but not any way they please. The fact that they "make" science means the presence of social factors is irreducible; the fact that it is "not any way they please" means invariants are involved. A principal task of contemporary philosophy of science, in my view, is to create a model of this interaction. In the absence of such a model, social constructivists will continue to stress the impact of social factors on science, while scientists and positivists will continue to press the point that scientific knowledge is independent of social and historical context. The argument will continue without resolution. Any adequate account of experimentation necessarily involves an adequate conception of this dual character—of how this antinomic character of scientific knowledge works. Until then, much of the work of social constructivists and others will continue to be perceived as a threat to science, as undermining the authority of science, and as contributing to skepticism about its value. Until then, the practice of science will continue to fall short of the false image promoted by positivist-inspired philosophers, leading to disillusion. Once such a conception is achieved, it will be possible to see the fundamental distinction, denied by many philosophers, between science, or the activity of disclosing truths about nature, and technology, or the activity of appropriating such truths for social ends outside the laboratory. This distinction, in turn, is crucial to other issues, such as the question of the ethical responsibility of scientists; on a more pragmatic level, it is also relevant to government funding practices. If there is no clear distinction between the activity of disclosing truths about nature and appropriating them for social ends, then governments are justified in de-coupling funding for basic and applied research programs and empha-sizing the latter over the former. Those who ignore these issues, and remain satisfied with conventional accounts of scientific practice, may be-come unwitting accomplices to the obstruction of scientific practice. As the authors of the *Nature* article bluntly put it, "the present financial crisis is to a considerable extent self-inflicted."[19]

Value of the Inquiry for Philosophy

Professional curiosity might lead philosophers to turn their attention to experimentation, just as professional curiosity often leads scientists to develop accounts of different types of natural phenomena for their own sake. But more is at stake here than professional propriety. The value of a rigorous inquiry into experimentation for philosophers can be seen upon considering the kinds of issues that such an inquiry needs to address. It would allow experimentation to be seen against the background of other human activities, and would draw the appropriate connections with per-spectives taken by the philosophical tradition on these matters. The in-quiry would encompass, at least in outline, the planning, executing, and witnessing of experiments, and to show how the issues involved are similar to or different from those involved in planning, executing, and witnessing other kinds of actions. Among the issues that it would be expected to cover are: the factors involved in the selection of which experiments to perform; the existence of a range of different kinds of experimental research pro-grams; the need for a specific site for the experiment and the varying desirability of different sites; the structure of teamwork and the collabora-tive nature of science; the existence and nature of artistry in experimental craft; what makes for a "good run"; the telling of good data from bad; the role of skill in the performance of an experiment; how laboratory skills are passed from the laboratory into the world; the relation between laboratory activity and technology; the way a scientific phenomenon can show up in vastly different environments and instrumental contexts and yet be said to be "the same"; how a hitherto unknown phenomenon is discovered or "recognized"; the possibility of self-deception and its difference from fraud; the nature of measurement; the relation between theories and ex-perimental acts based on and related to them; the nature of theory; the role of mathematics in theory; whether theory represents the world or anything at all; the "meaning" of quantum mechanics; the role of social and historical forces in science; the relation between science as inquiry and science as cultural practice; the writing of journal articles; the presentation of one's work at conferences; the writing of narratives about science.

Not all of these are *philosophical* questions. A small number, mainly those related to theory, have been broached by the traditional philosophy of science. A few more have been raised by the new initiatives into experimentation mentioned above. Still others may be better addressed by sociologists, historians, anthropologists, or psychologists. What is preeminently a philosophical task, however, is the attempt to construct a comprehensive view of experimental activity in which such questions are assigned their proper places to determine their relations to other questions, their incorporation within a conception of experimentation as a unified phenomenon, and their ties to existing philosophical perspectives. A philosophical approach to experimentation thus might be described as the attempt to construct a topology of experimental activity—a coordination of features of experimentation with the appropriate literature, philosophical and otherwise.

I do not claim to have such a topology fully worked out, nor to have complete answers to all these questions. I do think, however, that with the aid of the theatrical analogy it is possible to construct such a topology. The need for the theatrical analogy comes from the failure of traditional philosophy of science to break free of habit and adopt a sufficiently new perspective on its subject matter to reconstruct its own outdated topology. The theatrical analogy is not offered as part of a critique of science, but as an attempt toward reconstructing the language of the philosophy of science so that it can better understand and address its subject matter.

Proper recognition of the nature of experimentation in science thus will require more than a touchup in the traditional picture; it will force us to rethink our more general understanding of science. In addition, we shall have to consider the implications for philosophy itself. This neglect is so thorough and far-reaching that it hardly can be regarded as a mere oversight. True philosophical thinking involves not merely criticism but also coming to understand why one has been led astray. What the neglect of experiment reveals about the nature and direction of the philosophical tradition may result in the disclosure of forces that, unnoticed, have shaped that tradition, forces that have directed philosophers' attentions to certain areas and not others. That neglect, I shall argue, has been brought about by an ontological priority accorded to theory, and by the assumption that scientific knowledge is knowledge of the Demiurge, above human history and culture. But if the problem of experimentation is taken seriously, no such priority of theory or knowledge above human history and culture is possible, for scientific practice must always remain within the realm of human history and culture and maintain an irreducible tie to human practices even as it discloses phenomena with some degree of independence of any particular historical and cultural context. The study of experimentation will thus reveal something about the nature and direction of philosophy.

Genuine philosophy of science, far from being autonomous, is an emblematic branch of the discipline, incorporating within itself perspectives on virtually all aspects of philosophy including aesthetics. To appropriate a remark John Dewey once made about aesthetics, the greatest test of whether a philosophical system has a sufficiently adequate, flexible, and extensive grasp of experience lies in its philosophy of science. In its encounter with scientific activity, a philosophy is called upon to exhibit how most of the critical philosophical concepts—including truth, action, ontology, epistemology, praxis, poiesis, performance, production, recognition, phenomenon, aesthetics, abstraction, inquiry, invariance, interpretation, historicity, narrative, and theory—engage the world. This encounter also reveals the *spirit* of a philosophy. Does it shun science out of fear that philosophical knowledge suffers in contrast with scientific knowledge? Or does it debase itself in another way, by declaring without further ado or examination a particular model it has of scientific activity to be the paradigm for all human intellectual inquiry, including philosophy itself? Does the philosophy conceive of knowledge, scientific and otherwise, as amounting to the control and domination of nature and circumstance? Does it deny or downplay the role of theory and mathematics in areas where they are indubitably present, fearing that to admit the fundamental role of these in human activity would amount to a turn away from experience? Or does it conceive of the possibility of an artistic relation to nature— artistic even in the presence of a fundamental tie to theory and mathematics!—in which experimental activity amounts to the play of prudence and artistry in bringing forth natural phenomena, an activity that delights in the novel and surprising and can be accompanied by inspiration and the enchantment of creation? Its account of science thus prefigures a philosophical system's conception of the relation between human beings and nature, as well as the joint operation of an ensemble of philosophical terms. Our philosophy of science is a test of how seriously we take philosophy itself.

The absence of an existing coherent framework in which to discuss the broad range of issues involved in experimentation demonstrates that contemporary philosophy has largely failed this test. Traditional philosophers of science have avoided questions involving the actual practice of science in favor of a set of problems, mostly logical and epistemological, that have arisen within their own philosophical traditions. Topics such as the "logic" of induction, the "logic" of confirmation, the "logic" of discovery, Bayes's theorem, and so forth, with which certain philosophers of science are often occupied, do not concern scientists in their day-to-day work. But neither is science intrinsically concerned with dominating and controlling nature through abstractions that bear no relation to human time and history, as other philosophers have held. Science is broader in scope and more complicated than the fanciful, one-dimensional way in which it generally has been depicted by philosophers.

One may well question the validity of my proposed solution; one may question whether my extended analogy between scientific experimentation and theatrical performance is well-founded and useful given a sufficiently abstract conception of performance, or merely poetically suggestive. The answer will depend on what appears or is disclosed in the performance, as it were, that follows. But more is at stake than justification of the value of the theatrical analogy. I have already suggested that the accusations made against the philosophy of science mentioned at the beginning of this book are largely valid if one is speaking of traditional philosophy of science, and that they could and should be further developed and elaborated. What this book aims to do is show the kind of thing that genuine philosophy of science does. My inquiry takes the form of an invitation to look at the performance it enacts and to judge for yourselves the value of what appears. A more legitimate form of defense is inconceivable.

My use of the theatrical analogy can be viewed as a kind of experiment itself. It engages with the subject to sketch out in advance the kinds of aspects we should inquire into, and the kinds of outcomes that would satisfy us. It requires a certain training to apply fruitfully. It is not definitive, but always open to reevaluation and reexecution. Finally, as in the case of any truly consequential experiment, if it raises more questions than it answers, so much the better.

I

THE MYTHIC ACCOUNT OF EXPERIMENTATION

It is now commonplace in books and articles by contemporary philosophers of science to proclaim the discipline to be in a state of crisis.[1] Canonical authors such as Carnap, Popper, Russell, and even Quine have lost
their authority; few researchers are sure which way to turn; a variety of
modest initiatives are proceeding in a number of different directions.
Throughout this period of crisis (which may have lasted from one to four
decades, depending upon the diagnosis), philosophy of science has
seemed like a diseased physician who struggles futilely against a raging,
self-inflicted malady. The physician is so consumed by the struggle that
hardly any time is left for a glance at the "patient," science itself, who
meanwhile is managing quite nicely, utterly oblivious to and unaffected
by the paralysis of its alleged general practitioner.

The Mythic Account

The self-inflicted malady is partly due to the failure to adopt a critical
enough stance towards a certain ready-to-hand, conventional, and even
what one might call a *mythic* account of science and experimentation that
has such a powerful and pervasive influence on our conceptions about
science as to effectively foreclose questioning about it. Like other myths,
this one articulates a certain understanding of the deep structure of a
particular area of human experience. Also like other myths, this one is so
comprehensive that satisfactory answers (within the terms of the myth) to
nearly all questions can be found within it, as well as the rationale for
ongoing activity. We must become aware of the presence and scope of
this mythic account before proceeding further, in order to appreciate how
thoroughly one should be suspicious of conventional accounts of experimentation; moreover, unless myths are exposed, one tends to feed on
them, however unconsciously.

The myth addresses a division of labor present in modern science between experimenters and theorists. The distinction is most prominent in

physics, where contemporary practitioners of each group belong, in effect, to separate cultures. The division extends to the most mundane levels; when physicists at the Stanford Linear Accelerator Facility hold their annual softball game, there is no need for lengthy choosing of sides—it is "Theorists" versus "Experimenters." (The latter generally prevail.) Even in those fields in which individuals may not be so easily categorizable, such as geology or microbiology, the activity of science involves distinguishable theoretical and experimental roles.

But what is the meaning of this division? What are the separate roles that theorists and experimenters play, and how are they related? The mythic account provides a two-part answer. First, science is essentially theory making. Theories, in this view, are representations of the nature and behavior of a set of fundamental things that are "out there" in the world; true theories are accurate representations. These fundamental things can be material (in what might be called the Democritean version of the myth), formal (in what might be called the Platonic version), or unknowable (in the Kantian version).[2] In all three versions the basic content of science is removed from sense perception: the Democritean by the assumption that the fundamental entities are too small to be perceptible; the Platonic by the assumption that the fundamental entities are akin to ideal forms, numbers, or mathematical objects; and the Kantian by the assumption that the fundamental entities are unknowable things-in-themselves that are revealed in human experience only via the mediation of a synthetic activity that simultaneously puts them behind an impassable barrier. In representing the nature and behavior of fundamental things or their effects, theories explain observations—what we perceive—of the world. The truths aimed at by theoretical representations are held to be eternal, above human time and history. Experimental methods and practices, which are historically and culturally bound, play no constitutive role in them.

Second, experimentation is essentially theory testing. Not only is there an ahistorical structure "out there" in the world in some real or ideal form, but in the process of experimentation one can check what is and is not an accurate representation of that structure without affecting what is represented in turn. Experiments are not to be thought of as creations, as constituting or making things, as bringing things into the world, but as tests judging the correlation between representations and structure represented. Experiments aim to confront the experimenter with clear and distinct answers to questions, and are diaphanous with respect to the answers. To return to Kant's judge-and-witness metaphor mentioned in the previous chapter, it is as if the witness came to the stand already in possession of answers to the questions posed by the judge—already equipped with true statements about the world—and all the questioning process does is to provide the occasion on which they come forward. Experimentation is thus conceived as an essentially automatic process involving a minimum of interpretation; the witness is reduced to the status of the video cubes and related equipment of the "Concentration" game. Data, the product of the experimental process, is understood in this view to consist of fixed, eternal, and unproblematic facts.

The mythic view envisions science as taking place via a reciprocal inter-action between theory and experiments, questions and answers. But the mythic view also grants a clear priority to the role of theory. Theorists emerge as akin to visionaries, for they have insights into the basic structure of the world, while experimenters perform the less romantic task of deter-mining whether these insights are accurate, or of providing new clues for further theories. Experiments do not play a constitutive role in what theo-ries are about. The myth of experimentation is thus linked with a general view of science as principally a body of information rather than practices.

A classic example of the myth in action is the traditional account of the most famous experiment of all, that performed by Galileo at the Leaning Tower of Pisa.[3] Every schoolchild knows the story: Galileo came up with a hypothesis (that gravitation attracts different masses equally), climbed the Leaning Tower, and dropped a pair of differently weighted balls simul-taneously. Let us notice the minimum account of action that took place in this alleged episode: releasing two balls from one's hands and watching what transpired a few seconds later. The balls, in the schoolchild account, struck the ground at approximately the same time with a slight difference due to air resistance, thus refuting Aristotelian mechanics at a stroke and laying the cornerstone for modern science by illustrating the method of subjecting theories to experimental adjudication.

The account, as historians have shown, is largely mythical. Serious doubts exist whether Galileo ever performed the experiment.[4] Even if it took place, almost certainly it would have been inconclusive. Galileo's theo-ries called for the two balls to land at the same time with a slight difference due to air resistance, while Aristotelians would have expected the heavier ball to land considerably ahead. A modern reconstruction of the alleged event using a shotput and a softball has shown that the air resistance would have caused the former to land a full twenty to thirty feet in front of the latter, somewhere in the middle of the two predictions.[5] Both Galileo and the Aristotelians therefore could have claimed victory; at the very least, the experiment would have provoked heated debate over how to interpret the outcome.

Why, then, does the story seem like cultural bedrock to us? The answer lies in its mythological function; it is the scientific equivalent of the legend of young George Washington and the cherry tree, one that we know is almost certainly false but which we cannot refrain from passing along to our young. It forges a normative propaedeutic attitude to science; it in-structs the public and neophytes as to what we think scientific procedure is all about. The mythic version of the Galileo story informs us of the supposed truth that whereas Aristotelian science relied on the authority of books and wise men to impose its theory of the world—an irrational decision-making procedure—the new, Galilean science based itself on the true, rational, nonauthoritative decision-making procedure of experimentation. But the rise of modern science did not, in fact, mean the expunging of authority. Rather, it meant the development of a new tradition of looking at the world, different from and incommensurable with the old, in which authority, far from disappearing, merely changed form. It now appeared in the form not of privileged texts and teachers but of a complex hermeneutic involving methods and practices into which one is initiated when one studies science and through which the world is known—methods and practices that are continually being handed down, refined, and replaced.

Like other myths, this one has a theological origin in the notion that the ahistorical structure that theory attempts to represent has been inscribed in the world by God Himself. The classical image here was expressed by Galileo, who wrote that "the glory and greatness of Almighty God are marvelously discerned in all his works and divinely read in the open book of heaven," and that this book "is written in the language of mathematics."[6] This image—science as the reading of the book of Nature—accommodated several characteristic traits of thinkers of the Western intellectual tradition, including their discomfort with the opacity of nature, preference for the perspicuity of theory, and assumption that things are governed by an independently existing logic. Science, in the mythic view, allows us to see nature the way a divine Demiurge—the Author of nature—does; science, in short, is Demiurgic knowledge.

The notion that studying nature amounted to studying God through his craft predated Galileo. It is reflected in Biblical passages, such as the first verse of Psalm 19 ("The heavens declare the glory of God; and the firmament sheweth his handywork."), and by the ancient view that the heavenly orbs were gods and that elements of the divine permeated the natural world. Galileo's contribution was the conception of the contents of the book of nature as mathematical in character, with the corollary that its truths, once apprehended, were demonstrable with as much certainty as those of geometry. This contribution, however, changed the impulse of the image; in it, the truths of nature were depicted as encoded rather than openly revealed. The image was conservative in that it hearkened back to scripture and to Church doctrine and was framed in a language borrowed from Galileo's theological opponents. It was simultaneously progressive in the way it directed attention toward nature as possessing invariants of a mathematical character able to be discovered independently of scripture. But the image concealed as much as it revealed, for it directed attention away from experience and perception. In the modern version of this myth, science allows us to rise above our merely human concerns and perspectives—our history and culture—to become spectators to the "fundamental furniture"

of the Universe, and in its various versions this furniture is conceived as
being Democritean, Platonic, or Kantian; real, ideal, or unknowable.[7] But
this is still only a secularized version that derives from and is supported
by the deepest theological and metaphysical tradition in the West, that of
the vision of a divine *nous* contemplating a spectacle of *eidoi.*

The image also created and disguised a problem with the relation between theory and the world. If science is a reading of mathematical signs,
what relation do they have to perceptible objects and events? How can
experiments—real actions in the world—confirm or disconfirm ideal and
abstract symbol systems? It was possible for a time to conceive of mathematics as related to the world through *limit-shapes,* with measurement conceived as the infinitely perfectible process by which the limit-shape is
determined. Advances in technology appeared to provide a model for such
an infinitely perfectible process. The legacy of this image continued in the
tendency to treat science as theory, and theory as abstraction.

Aspects of this mythic account may sound at first obviously and unques-
tionably true, because of the way it is disseminated and reinforced in textbooks, children's literature, newspaper accounts of scientific activity, and by
many science historians. This account is also the way scientists themselves
habitually explain their activity to nonscientists, even when they suspect
its mythic aspect.[8] It is no surprise to find that theorists—who tend more
than experimenters to write books about science and who thus exert
greater influence on its perception by the public—frequently interpret their
own activity in light of this myth, affirm that the end of science is theory
production, and argue that the role of experiment is to support and guide
these efforts. Science, theorists tend to claim, aims at a purely cognitive
knowing or describing of the world. (This position might be contrasted,
say, with one that holds that the aim of science is to achieve a practical
ability to manipulate nature.)

Einstein, for instance, wrote that "the supreme task of the physicist is
to arrive at those universal elementary laws from which the cosmos can
be built up by pure deduction," and liked to stress the primacy of imagination and ingenuity over logic for attaining this end.[9] Feynman spoke of the
world as "a great chess game being played by the gods, and we are observers of the game." He continued, "We do not know what the rules of the
game are; all we are allowed to do is to *watch* the playing. Of course, if we
watch long enough, we may eventually catch on to a few of the rules. *The
rules of the game* are what we mean by *fundamental physics.*"[10] Here again,
the observational, theoretical side of science is accorded priority. Stephen
Hawking expressed a similar cognitive, theory-dominant view of science
in his enormously popular book, *A Brief History of Time.* "The eventual
goal of science," Hawking wrote, "is to provide a single theory that describes the whole universe."[11]

Traces of the theological dimension of the myth are also present in the works of such authors. While denying a personal God, Einstein stated that "a conviction, akin to religious feeling, of the rationality or intelligibility of the world lies behind all scientific work of a higher order. This firm belief, a belief bound up with deep feeling, in a superior mind that reveals itself in the world of experience represents my conception of God."[12] Hawking ends his book by remarking that discovering a single theory to describe the entire universe would enable us to raise the question of why the universe exists. "If we find the answer to that," he concluded, "it would be the ultimate triumph of human reason—for then we would know the mind of God."[13]

Einstein, Feynman, and Hawking are theorists, and we should not be surprised at their predilection for theory. But even prominent experimenters sometimes echo the myth in writing about their work, accepting its subaltern status. Nobel laureate Leon Lederman jokes that the goal of physics is to provide an equation for the universe that you can write on a T-shirt. A more serious example is provided by the work of experimental high-energy physicist Allan Franklin, author of *The Neglect of Experiment* and *Experiment, Right or Wrong*. Franklin is aware of what he calls "the general neglect of experiment and the dominance of theory in the literature on the history and philosophy of science," as well as of a mythic treatment of experiment, for he notes that ambiguities in the experimental process have been overlooked and historical details rewritten or obliterated not only by historians and philosophers of science but also by scientists themselves, who all too frequently lack knowledge about the history of their field. "The point here," he wrote, "is that real, as opposed to mythological, experiments are rarely discussed, even when experiment is mentioned at all."[14] But Franklin's approach—the publishing of case studies of specific experiments—can do nothing to dispel the myth if it does not attempt to challenge whatever conception of science brought about the neglect of experimentation in the first place, and to uncover the causes that might have led to it. And Franklin winds up not only not challenging the orthodoxy, despite awareness of its insufficiency, but he even supports it. For instance, he sees the function of experiment as that of assisting scientists in theory making, which he, too, regards as the principal activity of science. Indeed, Franklin describes the first of the two primary aims of *The Neglect of Experiment* as to help shed light on the role that experiments can and should play in the choice between competing theories or in the confirmation of theories. "The reader should conclude," Franklin says at the end, "that experiment has a philosophically legitimate role in the choice between competing theories and in confirmation of theories or hypotheses"—a conclusion fully in accord with the myth, and one which he himself admits sounds a mite commonplace.

The myth has been both extended and undermined in our century: extended, by the appearance of novel candidates for the fundamental furniture of the universe—symmetry groups, gauge particles, quarks, preons (postulated quark components), to list a few—and undermined, by the introduction of mathematical forms into theories that render ambiguous the way theories speak about the world. The theory of relativity to some extent, and quantum mechanics to a much greater extent, cannot be construed, for reasons too frequently discussed to warrant mention here, as picturing states of a real phenomenon apart from specific measurement contexts. To question that myth and seek a new image of scientific activity is not meant to challenge the validity or significance of science as an intellectual achievement, but to attain a better understanding of that validity and significance.

Philosophers and Experimentation

Ordinarily, philosophers are the right kinds of individuals to consult regarding the validity of myth. They have charged themselves professionally with studying the nature and practice of human activities from the perspective of an educated and critical audience, and in so doing have helped sweep away the veils of illusion that inevitably surround fundamental activities in any culture. Philosophers typically begin by asking questions that seem superfluous due to the availability of apparently satisfactory answers; philosophers are, as it were, professional askers of "stupid" questions. Many times, however, answering such questions adequately proves to be quite difficult, for the concepts that seem at first to be simple and the activities that are second nature turn out to be quite complicated and to have been determined by hitherto unknown forces. Attempts to answer such questions thus can provide new insight into things we thought were obvious and can open new possibilities. "What is experimentation?" has the earmarks of just such a promising stupid question.

Why, then, have philosophers failed to raise the question of experimentation? In part this is due to a predilection toward theory, with a concomitant neglect of action, leaving philosophers without adequate tools to handle experimentation in a philosophically appropriate manner. The move away from action as a philosophical theme was practically contemporaneous with Western philosophy itself, and was at work when Plato began using the Greek word for knowledge, *episteme*, which for Homer and others had meant practical skill and ability, with an emphasis on its ideational content. Plato's successor, Aristotle, viewed philosophical activity as a striving for *episteme theoretike*, the disinterested contemplation of eternal and unchangeable things. The notion that genuine knowledge is cognition, and cognition the mind's beholding of fixed and immutable forms, became a recurring theme within the Western philosophical tradition.

John Dewey, one of the few major thinkers to challenge this view, attributed the origin and reason for its success to what he called a "quest for certainty." Far from representing a quest for truth, he said, the philosopher's penchant for unchanging abstract ideas is instead merely a reification of the deeply rooted psychological desire, held over from the species's more primitive and precarious times, to rise above the frustrations and perils of the world and seize hold of the permanent. The character of this quest dictates the character of what would satisfy it. The quest cannot be conducted in the practical realm, for "practical activity deals with individualized and unique situations, which are never exactly duplicable and about which, accordingly, no complete assurance is possible."[15] The quest for certainty is satisfiable in the mind alone, in the grasping of eternal and immutable facts about the world. It is thus more realizable in concepts than in acts, in theories than in experiments.

A first corollary of this view, according to Dewey, is that genuine acts of knowledge cannot be originative, for if they altered or created the known it would not be eternal and immutable. Instead, knowledge consists in the formulation of ideas that agree with some pre-existing state of affairs. Knowledge is thus necessarily a *discovery*, not a making, and what is discovered is antecedent to and unaffected by the inquiry that discovers it. The act of knowing is thus modeled on what supposedly transpires in vision; that when a person's gaze meets an object, the eye is independent of the object beheld and neither creates nor modifies it. A similar passivity is supposed to characterize the mind in knowing, and Dewey therefore referred to this version of what I have called the mythic view as the *spectator theory of knowledge*. A second corollary is that knowledge involving the practical realm—the world of the changing and temporal—is an inferior kind. This view therefore leads to the separation of the realm of knowledge and practical action and the elevation of the former over the latter. "Thus the predisposition of philosophy toward the universal, invariant and eternal was fixed. It remains the common possession of the entire classic philosophic tradition" (LW4:16).

Dewey was sufficiently conversant with contemporary scientific developments to realize that Einstein's work on relativity and that of Heisenberg and others on quantum mechanics revealed the spectator theory to be erroneous. "The principle of indeterminacy thus presents itself as the final step in the dislodgment of the old spectator theory of knowledge. It marks the acknowledgment, within scientific procedure itself, of the fact that knowing is one kind of interaction which goes on within the world" (LW4:163). Dewey spoke too soon in announcing the demise of the spectator theory, but he did have an accurate grasp of its flaws. It was hardly necessary to delve into the frontiers of basic science to reveal the limitations of the spectator theory, he wrote; it was sufficient to consider the patent successes of modern science and technology in controlling nature.

Dewey did not worry that any challenge of the spectator theory would undermine the status or authority of science, for, he felt, the "security attained by active control is to be more prized than certainty in theory" (LW4:29–30). The spectator theory was simply outmoded, in his eyes, and he wanted to find something to replace it. "Is it not time to revise the philosophical conceptions which are founded on a belief now proved to be false?" (LW4:85–86). To find a replacement, one had only to examine how the successes of modern science and technology had been achieved. This examination, Dewey says, reveals that knowledge does not result from human beings detaching themselves from the world or reproducing it in thought but from their active encounter with it. "Experimental procedure is one that installs doing as the heart of knowing" (LW4:29).

The neglect of experiment is a symptom of the more general neglect of action on the part of traditional philosophy. No interesting philosophical problems are thought to be present in experimentation, in the traditional view, because no interesting philosophical problems are thought to be present in action. The two corollaries of the quest for certainty cited above correspond to two aspects of the traditional philosophical account of science also mentioned previously. The first corollary, that genuine acts of knowledge are not originative, corresponds to the view of science as doing nothing more than reading an already written text, the book of Nature. The "Concentration" view of experiment, in other words, is but an expression of the spectator theory of knowledge. The second corollary, that knowledge involving the practical realm is an inferior kind, corresponds to the tendency to value theoretical over experimental activity. Hence the existence of a profound ambivalence on the part of traditional philosophers of science towards their supposed object of study. On the one hand, they have been enthusiastic about, embraced, and even worshipped the scientific enterprise. On the other hand, and equally resolutely, they have refused to see what it really involves. Monochromatically, philosophers of science have zeroed in on those aspects of science having to do with theory and overlooked those aspects having to do with experimentation: with planning, skillfully performing, and witnessing acts.

In experimentation, human beings use instruments to make something happen. The instruments can be as simple as clocks, measuring sticks, and notebooks, or as technologically sophisticated as particle accelerators, interferometers, and electrophoretic equipment. The number of human beings involved can range from one to thousands, and what happens may involve complex subatomic interactions, chemical reactions, the behavior of materials, or conscious human responses to particular situations. Experiments may involve a great deal of mathematics in their preparation and analysis, or virtually none. Sometimes experiments are conducted with elaborate, sometimes with minimal, expectations regarding their outcome. In each case, however, experimentation involves the use of instruments to perform an act. Even the regularity of the motions of the planets becomes evident and publicly accessible only through the act of measuring and recording their positions.

To be sure, this vague description does not distinguish scientific experimentation from other activities that use equipment. Carpenters use tools to make furniture, farmers use agricultural equipment to tend crops, manufacturers use machines to transform metal into various products, but these are not instances of experimentation. Still, this preliminary description characterizes what experimenters do to a first approximation, and therefore the domain in which philosophical inquiry must operate. Even if all the requisite equipment and materials are assembled in one place in the laboratory, an experiment is not yet underway; an experiment requires an active involvement of things with each other and of human beings with them. Even before the publication of journal articles and discovery announcements; even before theories are confirmed, refuted, or invented; even before colloquium talks, graphs, and numbers, there is a doing of something. What is primary in a scientific experiment is not a publication, observation sentence, or data table, but an *act*.[16] An appropriately structured philosophical inquiry into science can not therefore concern itself with theory alone. Nor can it assume that data simply exist. It must inquire into the nature of actions, which are prior to data; how such actions are planned, performed, and understood.

One must be suspicious, therefore, of the waves of enthusiasm for science, periodically displayed by philosophers at different moments in the history of the discipline, that have not broached the issue of action. In the early seventeenth century, for instance, many philosophers, looking with envy on the evident successes and tangible fruits of science, attempted to appropriate science as the only legitimate form of natural philosophy— some made it the paradigm for human intellectual inquiry in general— and sought to enlist themselves in its service. But this enthusiasm did not involve an inquiry into experimentation. René Descartes's handbook on scientific method (1628), for instance, proposes a set of rules, but not for experimentally engaging with nature; they are rules "for the direction of the human mind."[17]

The early twentieth century witnessed another renewal of the ancient quest for certainty and of widespread disillusionment with the philosophical tradition, and many philosophers once again looked to the sciences for guidance. Bertrand Russell, who once remarked that he wanted certainty in the kind of way in which people want religious faith, proposed in the final chapter of his influential 1914 work, *Our Knowledge of the External World, as a Field for Scientific Method in Philosophy,* that philosophy would progress only when it cast aside tradition and enlisted individuals with scientific training to tackle philosophical problems. A few years later, Ludwig Wittgenstein argued, in the *Tractatus Logico-Philosophicus,* that science discovered what was true about the world; philosophy only assisted by clarifying unclear statements and eliminating meaningless ones. More extreme still were the members of the Vienna Circle, who made of science the ultimate arbiter of truth and meaning and who turned philosophy into science's amanuensis; they entitled their 1919 manifesto *The Scientific Conception of the World.*[18]

In view of how formative their approach was for later directions in the philosophy of science, it is useful to recall just how extreme and messianic were the members of the Vienna Circle. According to its famous *verifiability principle*, a sentence is meaningful if and only if it expresses an analytic or synthetic truth; that is, if it is a scientifically verifiable empirical hypothesis or a logical tautology. All other sentences are meaningless. The job of philosophy was to analyze sentences to see whether they expressed synthetic truths, analytic truths, or were nonsense. The Vienna Circle members thus saw philosophy as carrying out a sort of triage of perplexing sentences, deciding which were to be assigned for clarification to scientists, which to mathematicians and logicians, and which to the dustbin. They viewed themselves as righteous champions of knowledge and the good who were combatting the forces of superstition and ignorance. Their adoption of the scientific spirit did not mean that they had abandoned philosophy's age-old attempt to address the problems of human civilization and destiny. To the contrary; philosophy, they argued, would *solve* them once and for all by becoming the servant of science, facilitating and promoting its entry into all aspects of human life. "The representatives of a scientific *Weltauffassung*," wrote Otto Neurath, a leading figure of the Circle, ". . . know only science and the clarification of scientific methods, and this clarification is all that remains of old-fashioned 'philosophizing'. . . . What can not be regarded as unified science must be accepted as poetry or fiction." And Neurath and two other leaders of the Circle flatly declared, in a manifesto of 1929 entitled *The Scientific Conception of the World: The Vienna Circle,* that "The scientific world-conception knows *no unsolvable riddle.*"[19]

The science championed by Neurath and his colleagues was that of the mythic account. Experimentation meant verification or confirmation, which were viewed as basically matters of simple observation; experimentation was but observation by other means.[20] Scientists make observations—such as "here now pointer at 5, simultaneously spark and explosion, then smell of ozone there"—and then decide whether they confirm the theory in question.[21] Verification or confirmation is thus a matter of the relation between two sets of *sentences,* one consisting of "observation," "protocol," or "basis" sentences, the other of "hypothesis" or "theory" sentences. Whatever it is that gives birth to the statements of the observation language evidently was not an issue for them.[22] This much is clear from the characteristic examples used by members of the Vienna Circle. A recurrent one is: To what extent is the hypothesis-sentence "All ravens are black" confirmed by the observation-sentence "*a* is a raven and *a* is black"? To be sure, confirmation was a hotly debated issue for the Vienna Circle, and has been for philosophers ever since. But awareness of the complexities of confirmation—a logical issue—is not equivalent to awareness of the nature of experimentation. That executing and understanding an experiment might be fundamentally a more complicated activity than determining the color of a black bird does not seem to have been explored by these proponents of "the scientific conception of the world."

The science-worship, or *scientism,* of the Vienna Circle thus did not originate in a critique of science for its own sake; it originated elsewhere, in the reinvigorated quest for certainty and the desire to demolish traditional metaphysics. (The social, historical, and even psychological motives that conditioned this invigoration need not be discussed here.) In implementing this quest, the Vienna Circle insisted that all true knowledge was to be grounded in facts in the way that they assumed was the case in science, and they spun their doctrines from this assumption. But they never made it their business to examine critically how knowledge was in fact obtained in scientific activity.

Although the Vienna Circle dissolved in the late 1930s, its spirit continued to flourish, particularly in the United States, where the positivist style underwent a shift of direction known as the *linguistic turn.* Philosophers realized that the meaning of sentences depended not only on whether the terms contained in them referred to things in the world, but also on the syntactical structures of language; that is, on the interrelationships of terms and expressions. They continued to categorize sentences into tautologies and scientifically verifiable empirical hypotheses, but the tautologies concerned were a priori structures of language—structures that, when projected upon the world, make it intelligible. Attention turned more and more to these structures, to the role they play in thought, and to the process by which we choose them—to, in short, the conceptual structures by which objects and events in the world are grasped rather than to the actions and performances that give rise to these objects and events. Throughout the postwar era, science never lost its robust influence on mainstream American philosophy. "[I]n the dimension of describing and explaining the world," wrote University of Pittsburgh philosopher Wilfred Sellars in an influential book of the 1960s, "science is the measure of all things."[23] Yet experimentation, the way scientists literally "take the measure" of things, was thematized neither by Sellars nor by anyone else. Philosophers of science have largely pursued a set of logical problems—such as the nature of confirmation, induction, and probability—that have arisen within their own traditions, and they have simply assumed that the optimal solutions to such problems are the normative methods of science itself as the most rigorous and obviously successful way of acquiring knowledge about the world.

Consider, for instance, the way in which the problem of experimentation manages to be overlooked within both empiricism and its antithesis, anti-empiricism. The empiricist perspective conceives of facts or observations as the building blocks with which science begins, and theories as ways of organizing these building blocks, arrived at by some process of generalization or induction. More recent empiricists have repudiated the naïveté of their intellectual predecessors' belief in the possibility of an observation language, but they still tend to emphasize the existence of a solid empirical base that accumulates and remains the same through theory change (and that retains a priority in the structure of science), without paying sufficient attention to the experimental acts by which that base was established. Even Ian Hacking, who in his admirable book *Representing and Intervening* points out complexities of experimentation frequently overlooked by philosophers of science, does not manage to incorporate his insights into a perspective challenging the one that led to the oversight.[24]

The other, anti-empiricist perspective stresses the contribution of theory to data, and that deciding between competing hypotheses or theories cannot be a matter of observation alone. Carnap ultimately adopted a variant of this view, rejecting the belief in foundational data in science and proposing that the adoption of a theoretical framework was required before observation sentences could be formulated.[25] Many who believe that facts are "theory-laden," to use the term promoted by philosopher N. R. Hanson, maintain the existence of a stable, factual level of some sort in experience, and that what varies is the conceptual framework in which that level is apprehended. For Karl Popper, "the empirical basis of objective science has thus nothing 'absolute' about it. Science does not rest upon solid bedrock. The bold structure of its theories rises, as it were, above a swamp."[26] But the swampiness that Popper sees is not due to the recognition of any ambiguities in experimentation, only to his belief that the meaning of observation statements depended on the theoretical structures used to formulate them. As he writes, "observations, and even more so obversation statements and statements of experimental results, are always *interpretations* of the facts observed . . . they are *interpretations in the light of theories.*"[27] W. V. O Quine, too, has a faith in the stability of observational language; any "relativity" is due to the existence of different "webs of belief" with which we think about the world.[28] Experimental data form a swamp only because of the various conceptual frameworks with which the data is constituted. Choose a single framework, and the swamp turns for a moment into bedrock. Once again, the emphasis is on the choice of framework rather than the performing and witnessing of actions.

Far from probing beneath the *mythic treatment* of experimentation, philosophers have wound up sanctifying it. The *problem of experimentation* still has not arisen as an explicit philosophical theme even in the hands of individuals who mean to focus attention on it. It is not that the problem of experimentation has been swept under the rug; the problem has not even been noticed. Philosophers of science hitherto have acted as though they had adequately covered the experimental side of science by simply asserting that scientific theories are based on evidence; they then believed themselves entitled to move on to their real interest, the logical relations between theories and supporting statements. Nothing more need be said about experimentation, was the thought, because nothing more *could* be said. Even today, philosophers of science tend to assume that scientists go into laboratories, use whatever equipment they use with whatever skills they have—sometimes, to be sure, stumbling across things serendipitously and occasionally taking missteps—and come out again with fully formed, objective information about the world. Only at that point, according to the traditional account, do truly interesting philosophical questions arise.

The point of recounting the flaws in previous philosophies of science is not to belabor the sins of the past; it is to point out the deeply ingrained philosophical tendency to neglect experimentation and the kinds of stratagems involved. It will not suffice to add a few lines here and there to existing philosophy of science textbooks, nor to contribute a new chapter ("Experimentation") that will readily mesh with the approaches, issues, and concepts of the rest of the text. We must find a new way of discussing experimentation that approaches it as a *process* for which data and theories are not foundations but outcomes. It is my contention that the theatrical analogy—experimentation as the production and evaluation of performances—is the right tool for generating a model for this kind of process.

Philosophical Tools Needed

At the beginning of this chapter, I compared current philosophy of science to a diseased physician struggling futilely against a raging, self-inflicted malady so severe that the poor physician has not the will to attend to the would-be patient, who by contrast is in sound health. To recommence practice, two things must happen: first, the physician must self-administer the right therapy; second, the physician must begin to look again at the patient. The current crisis in the philosophy of science will be cured only with the advent of two things: the right set of tools, and a real look at science.

Scrutiny of the practice of science is necessary because the mythical account has such firm hold of our minds that we have become blinded to how science actually works and are thus estranged from the possibility of understanding it. Even fully drawing out the implications of laboratory lunch table anecdotes, and of items culled from notes and news sections of science magazines, may prove more disclosive than entire libraries of volumes written by philosophers. It might seem unprofitable for philosophers to pay attention to the day-to-day aspects of scientific activity, given the radically disparate backgrounds of experimental scientists and philosophers. Experimental activity involves a practical engagement with materials, techniques, and instruments in the production of actions, whereas philosophical activity involves a different kind of engagement in scholarly research in the service of speculative or systematic thought. We are from "two cultures," as C. P. Snow said.[29] And it is certainly true that the bulk of the working experience of scientists and philosophers is markedly different. But philosophy does not stand alongside other human activities as one *fach* (academic discipline) to another. Especially when it is a question of the "philosophy *of*" something, philosophers need to draw on a certain involvement with or at least appreciation for their subject matter.

The great temptation to which philosophers often succumb is to empty their bookshelves, sweeping away whole bodies of texts or areas of experience as not requiring their attention. Traditional philosophers of science have fallen into this temptation regarding the practice of science. But involvement is a precondition for understanding, and philosophers of science need to engage themselves in how science works. Nothing must be ruled out ahead of time as too trivial, for one never knows which details will prove disclosive. One might find oneself looking into areas that according to the traditional approach belong to the backwaters of science. Einstein once wrote that the physicist cannot surrender to the philosopher the task of critically contemplating science, "for, he himself knows best, and feels more surely where the shoe pinches."[30] Such was Einstein's understated way of saying that philosophers of science have tended to misconstrue what they see. If philosophers are not to surrender to scientists the task of carrying out the philosophy of science, they must begin by judging their work against the standards appropriate to it; the practice of science itself. To paraphrase Martin Heidegger, to do otherwise is like seeking to estimate the nature and powers of a fish by seeing how well it is able to live on dry land. For too long, we philosophers of science have been sitting on dry land; to help our philosophy, and spare our nostrils, we need to become more aware of the practice of science.[31]

A second requirement in dealing with our "patient" is the right set of tools and methods. Philosophers of science need to consult different kinds of texts than those considered authoritative in the past. Like scientists, philosophers stand on the shoulders of predecessors (whether they know it or not) by their knowledge of already developed tools, by their instruction in the ways in which the tools are used, and by their expectations of the ends towards which the tools can and ought to be put.

But the right set of tools for a task depends on the work to be done.

From the discussion thus far, we are already acquainted with several needs to be satisfied by an investigation into experimentation: these include accounts of inquiry, of invariant, and of interpretation. A brief review of what we know so far of these needs will provide basic clues for where to turn for tools.

An account of *inquiry* is required in view of the question that arose previously, concerning the role of practical activity in an inquiry whose end is knowledge. In experimentation, certain questions prompt human beings to engage themselves with the world to produce actions under the expectation that what'appears in those actions may lead to answers to the questions. But an action is a unique event, a particular physical presence, while the knowledge about the world sought after in an inquiry presumably takes the form of a universal. What part, then, can the production of actions play in an inquiry? The reply that knowledge takes the form of theories or models that predict actions of certain kinds in certain conditions merely shifts the problem, which now reappears in describing how an action fulfills or fails to fulfill a theory or model.

An account of *invariance* is required because scientific experiments have what I called the antinomic property of being able to be both performances of unique events in the world and at the same time performances involving appearances of phenomena (electrons, say) that exist or *are* above and beyond the specific experimental context. What is the nature of this invariance? Different answers are provided by different versions of the myth. In the Democritean version, what is scientifically most real and permanent are imperceptible atoms and the void; descriptions of the behavior of observable aggregates are but inventions by human beings to "read the surface" in a stable, local context of something that is deeply lawless. In the Platonic version, what is scientifically most real—indeed, what exhausts the content of genuine knowledge of any kind—are the eternal forms, which we uncover in the course of inquiry, while matter changes constantly. In the Kantian version, the presence of systematic interconnections in nature is neither the result of a contribution by the human mind to a set of already existing things in nature, nor the result of a preexisting, ontological structure that we subsequently discover; the existence of systematic interconnections is projected into nature by the spontaneous activity of reason in order that there be any experience of nature at all. "Reason . . . does not here beg but command."[32]

The issue can be dramatized by recalling the famous lamp hanging in the Baptistery in the Duomo of Pisa, whose swing, according to legend, inspired Galileo to consult his pulse and discover its isochrony—one of the first scientific invariants to be given mathematical expression. This is the pendulum that seemed both proletarian and patrician to Mark Twain, arousing in him feelings of awe and intimacy:

It looked like an insignificant thing to have conferred upon the world of science and mechanics such a mighty extension of their dominions as it has. Pondering, in its suggestive presence, I seemed to see a crazy universe of swinging disks, the toiling children of this sedate parent. He appeared to have an intelligent expression about him of knowing that he was not a lamp at all; that he was a Pendulum; a pendulum disguised, for prodigious and inscrutable purposes of his own deep devising, and not a common pendulum either, but the old original patriarchal Pendulum—the Abraham Pendulum of the world.[33]

What is the status of the lamp's isochrony? In the Democritean version of the myth, what is there most primordially is a temporarily stable configuration of atoms, and the isochrony but a way we humans have of drawing resemblances between that configuration and what appear to us to be similar aggregates; the relation between "parent" and "toiling children" is superficial rather than familial. In the Platonic version, on the other hand, the isochrony is precisely that part of nature that is truly knowable: the lamp is "Abraham Pendulum" only because it was the first material object in which the form was recognized by humans. In the Kantian version, the isochrony, neither subjective nor ontologically grounded, is as real for us as nature itself. An account of experimentation needs to take a position on the status of invariants: Is it one of these three possibilities, or are there others?

An account of *interpretation* is required because the activity of experimentation is interpretive in several ways. An experimenter interprets in deciding what kind of experiment to design, how to design it, what collaborators to enlist, what theories to apply, how to apply them, where to carry out the experiment, what kinds of equipment and methods to use, what actions to perform in running the experiment, how to apply skills, how to decide whether the result is "real" or an artifact of the machine, how to know when the experiment is ended. But interpretation is also involved in scientific activity in a more fundamental way. Once one views knowledge as an act rather than as the content of an act, one places at the core of all knowledge, including scientific knowledge, an essential relation to human culture and history. For truth now involves a disclosure of something to someone, a disclosure that always takes place against a particular determinate cultural and historical context. Every action, every act of understanding, every disclosure of meaning becomes a matter of interpretation. Hence, there is range of different kinds of interpretation involved in experimentation that must be distinguished in any account of it.

Inquiry, invariant, and interpretation are of course interrelated; resolving many of the issues already mentioned (such as, for instance, developing a model that recognizes both the place of social factors and of invariants) requires taking a position on their relation.

In the next chapter, I shall outline existing philosophical perspectives that provide basic tools for an account of inquiry, invariant, and interpretation. These tools are but a starting point; they will need to be adapted for the context of experimentation. Nevertheless, they will provide a philosophical basis for the framework developed in the chapter that follows. Even in a philosophical account of a novel phenomenon one needs to ground one's work in elements of the philosophical tradition, for in philosophy as in science it is never possible to start entirely from scratch. A scholar enters a field already structured by those who have come before, and those structures create a context in which discussions of an orderly sort can take place, develop, and ultimately transform the field. Just as one would have no reason to trust a proposed new cancer remedy or epilepsy cure that had not been shown to have a basis in contemporary practice or which corrects that practice, so a convincing philosophical account depends on reliance on existing work, however much an inquirer may seek to transform it. Philosophy, too, needs to avoid the scholarly equivalent of cold fusion episodes, in which lack of care in tying one's work to existing practice produces what first appears to be a novel discovery of earth-shattering significance, but which later turns out to be sloppy work. Scholarship, like science, is innately conservative, respects existing language and procedures, and changes them only after reflection. In this way a path is built that, when followed, eventually leads to transformations of the subject matter. Even the treatment of a novel phenomenon begins with tools that are present in the tradition even when it intends to transform them. My introduction of these traditional accounts will prepare the way for the theatrical analogy in chapter 3. The theatrical analogy is not an attempt to start from scratch, but to liberate the imagination by pointing to a new set of philosophical tools that may be used to understand science, and to coordinate their use so that a coherent picture of it emerges.

II

PHILOSOPHERS AND PRODUCTIVE INQUIRY

Next to my bed is a bookcase of late-night reading material. Visitors usually think the order of books is random, but that is not so. One shelf, whose theme happens to be jazz dance of the 1930s, includes music books, autobiographies, poetry, movie and musical encyclopaedias, and literature about Harlem night clubs: in my inquiry into the subject I have discovered each to illuminate an important aspect. Another shelf includes *Hamlet* and *The Interpretation of Dreams;* its coherence becomes clear when one peruses the books in between on psychology and character. Out in my library, the organization is more conventional. In areas devoted to "philosophy" and "literature," I keep the appropriate books in alphabetical order. But next to my bed, the organization is improvisational. I find myself often straying from my original intentions, "filling in" shelves with no explicit rules, and inventing new categories on the spot; moreover, I often wind up pursuing directions that do not pan out, forcing me to reshelve things. *Hamlet* may wind up back amid plays or English literature. But what I am doing is serious and important; I am shaping new categories in addition to filling out existing ones. And this process is not fundamentally different from that used by the broader social community to order its books—the order reflected in my library. That community, too, is constantly "expanding its canon," adding and displacing books through inquiry.

The process of assembling a shelf of books is like the improvisational character of the growth of knowledge.[1] Human knowledge is akin to an extensive set of shelves, with the books on each shelf belonging together because they address a particular activity or feature of the world. Deciding which books to place on a shelf is an act of interpretation. If the tradition is well established, the act of interpretation need not be original; if not, an interpreter may need to demonstrate the rationale for each new book. Each shelf thus has an intelligibility; it represents a "tradition of interpretation" about some worldly phenomenon. But no matter how carefully we judge, material on one shelf may also be placed on others; a copy of *Hamlet* might well appropriately sit next to a work by Freud or on relations with parents. Suppose it is a book by Freud; that book may provoke questions about Hamlet's behavior best addressed by adding other works by Shakespeare, while questions raised by these responses may cause us to add other works by Freud, works of other playwrights, biographies, other kinds of art works, etc.

The ultimate undecidability of book shelving should be no surprise; the phenomenon occurs in science when results in one area unexpectedly become critical to another. Interactions between mathematics and physics are well known, but consider the recent surprise interaction between microbiology and knot theory, once a mathematical oddball. The interaction was created by the discovery that DNA is coiled up in tangles, so that understanding the replication process, involving the unsplicing and resplicing of strands, requires knowing the behavior of knots. And in 1900, who could have guessed that books on the idea of the quantum, originally proposed to solve problems in black-body radiation, ultimately would wind up belonging in libraries devoted to occupational safety, via a route leading from the quantization of energy to the quantization of magnetism to superconductivity and the Josephson Junction to SQUIDS (Super Quantum Interference Devices) to magnetopneumography, the study of magnetic fields given off by lung activity. Odd linkages similar to these are often found in the notes and news sections of science journals; the point is that the forging of novel connections between apparently disparate phenomena is not surprising but mundane.

Maintaining a shelf is thus an ongoing process of inquiry that requires constantly adding and subtracting books. Initial questions may turn out to have been vague, misdirected, or insufficiently focused, or they may simply lead to different questions. Not every shelf is successful. Inconsistencies develop between books or groups of books on a shelf—as is the case, in my opinion, when people place books about quantum mechanics on the same shelf with books about Eastern mysticism, and attempt to make them address one another.[2] In successful inquiry, assembling the new shelf allows us to see the thing addressed in a new light and causes us to rethink the meaning of books already there. ("Seeing" is here used as a metaphor for perception in its widest sense.)

I have already indicated that the shelf labeled "experiment" in the philosopher's library is almost bare or at best chaotic. There is no tradition of inquiry. The initial choice of books is thus not *objectively* determinable, and depends on the inquirer's finite knowledge of the library and judgment about the phenomenon. Nonetheless, lack of an explicit set of rules does not mean an action lacks constraints or is the arbitrary outcome of a set of merely private considerations. The inquirer composing the bookshelf does so to answer questions interesting a community of readers about a subject. The inquiry therefore addresses expectations concerning certain problematic issues, meaning that the initial choice of books is socially negotiated with a public dimension. It must be possible for the community of readers to rediscover the meanings intended by the inquirer (whence the reason the narrative voice of many philosophical works occasionally lapses into the first person plural). If that rediscovery eventually takes place, the inquiry is meaningful no matter how idiosyncratic the choice of first books once may have seemed; ultimately a convergence occurs as the object comes ever more into view.

At the end of the last chapter, I mentioned three problematic issues pertaining to experimentation: inquiry, invariant, and interpretation. An extensive philosophical literature exists about each, with certain authors— John Dewey, Edmund Husserl, and Martin Heidegger, respectively—surfacing as especially innovative and insightful. This provides us with a natural choice of first books; once on the shelf, they allow an initial perspective on experimentation and enable us to begin the process of inquiry. For philosophical truths do not pop out of texts any more than scientific discoveries pop out of equipment; a process of "tuning" is inevitably required as one's eyes become adjusted to seeing something anew, and, as one checks to make sure one is seeing aright, leading one to pose new kinds of questions for further inquiry. Spontaneous clarity is possible only with what is already understood. In philosophy, as in science, missteps that one takes in a course of inquiry will be pointed out and eventually corrected. Above all, one must preserve the spirit of free, unfettered, and unabashed questioning that is essential to the spirit of both philosophy and science. To do otherwise—to commit oneself to seeing things through to the end within established traditions—would result in certain short-term gains; one's work would be more readily recognized by colleagues as "philosophical." But that would also run the more significant danger of thwarting recognition of the genuinely new.

Inevitably, the ensuing summaries neglect or mispresent elements of their thought. But summaries are essential here, given the chaotic state of the shelf and the vastly different backgrounds and perspectives of potential inquirers. If my travesties are serious enough, those with deeper insights will be able to reveal this in turn.[3] One could then evaluate whether to reopen the inquiry from the beginning or whether, as I hope, the subsequent twists and turns of the inquiry might render a reopening unimportant. The outcome of this decision naturally would depend largely on the interests and questions motivating the new inquiry: whether, for instance, it aimed to determine primarily the nature of experimentation or the character of that particular philosopher's thought.

John Dewey and Inquiry

Dewey, a founder of the philosophical movement known as pragmatism, devoted much study to how experimental acts are planned and executed in inquiry. His efforts to develop a theory of inquiry were strongly shaped by the example of science; *The Quest for Certainty* is explicitly devoted to the study of the procedure of natural science. So important is Dewey in the American philosophical tradition, and so central are reflections about science to his thought, that any serious discussion of the nature of science that does not eventually confront this or other works of his is guilty of conscious avoidance. Yet Dewey has not been on the shelf of mainstream philosophy of science; he was displaced by the success of positivist-inspired and analytic philosophy, which purportedly brought greater rigor and precision to the subject. Dewey's works do lack the trappings of rigor; he is no formal system builder; his prose, somewhat disorganized and unstylized, contains little jargon and can be read by the interested lay person as readily as by the specialist; his examples are simple and nontechnical. Dewey's was a deeper rigor. He focussed, as few contemporaries did, on the fact that knowledge is the product rather than the content of human actions, and took seriously the implications. His books are thus natural candidates to set on a shelf about experimentation. In what follows, I do not provide an overview of Dewey's philosophy; my aim is not scholarly propriety but disclosure. I shall pick and choose from his works to suggest how one might obtain an initial perspective on inquiry adequate to a treatment of experimentation, and acquire in the process a rudimentary vocabulary and a sense for what other books might fruitfully be added to the shelf.[4]

Dewey's philosophy, as mentioned in chapter 1, was a protest against a particular version of the mythic view of science that he felt was the dominant philosophical tradition of the day and that he baptized the *spectator theory of knowledge*. He wrote, "Experimental knowledge is a mode of doing, and like all doing takes place at a time, in a place, and under specifiable conditions in connection with a definite problem" (LW4:82).

Dewey developed this insight by analogy with evolutionary biology, in which organism and environment are mutually engaged in a dynamic, ever-changing relation. Human beings, like other organisms, have needs, desires, and capabilities that inevitably clash with their surroundings, giving rise to what Dewey calls *problematic situations*. All organisms respond by trying to transform or *reconstruct* problematic situations, but human beings are different in the means (thought and instruments) used to produce this reconstruction. Reconstruction changes the environment, the organism, and the forms of its interaction with its surroundings. Dewey calls the process by which human beings employ thought in the transformation of problematic situations *inquiry* and says that knowledge is the outcome of inquiry, hence of a making rather than a discovery. Through inquiry, human beings realize a more assured, enriched, and deepened experience and engagement with the environment. "If we see that knowing is not the act of an outside spectator but of a participator inside the natural and social scene, then the true object of knowledge resides in the consequences of directed action" (LW4:157).

A relatively simple example of Dewey's method at work is an early, now-classic paper of 1896, "The Reflex Arc Concept in Psychology" (EW5:96–109), challenging the view, then dominant in psychology, that experience is analyzable into three basic parts: a sensory stimulus or sensation, a central activity or idea, and a motor discharge or act proper—the three comprising what was known as the "reflex arc." For instance, Dewey wrote, the reflex arc could be used to characterize a child's experience in reaching out to a candle, being singed, and then withdrawing the hand. First, the child receives a sensation of light. That sensation provokes a certain idea in the child, which in turn initiates a response—the forward movement of the hand toward the candle. The outcome of that sequence of stimulus and response is a new stimulus (the burning sensation) followed by another response, the withdrawal of the hand. The reality of the event is thus seen as a mechanical sequence of distinct steps: "sensation-followed-by-idea-followed-by-movement" (EW5:97).

But for Dewey, every experience, no matter how simple, is an activity whose aspects are not discrete elements but phases of a whole.[5] Dewey views the initial state, for instance, not as a stimulus or sensation, but as a *looking*, which is a coordinated act in itself; in looking at the candle rather than at something else, the child does not passively encounter a sensation but actively attends to it—positioning the head a certain way, fixing the eyes, twisting the body, and so forth. And if the hand then reaches for the candle, it is not as a response to a stimulus, but another act that grows out of, enlarges on, and transforms the first. It, too, requires coordinated movement; hand, body, and eyes must cooperate in effective grasping. When the hand retracts upon being burned, it is not a response to the substitution of yet another sensation for a prior one, but the completion of the coordinated activity that transforms and enriches the experience of the child. Henceforth, when the child sees a candle, its experience is no longer what it was; "it is seeing-of-a-light-that-means-pain-when-contact-occurs" (EW5:98).

The application of the idea of the reflex arc may usefully represent experience in certain respects, Dewey says, but distorts it as well by the pretense that the elements are discrete rather than coordinated. "What we have is a circuit; not an arc or broken segment of a circle. This circuit is more truly termed organic than reflex, because the motor response determines the stimulus, just as truly as sensory stimulus determines movement." No doubt, Dewey says, someone will object that there is of course a distinction between stimulus and response. "Precisely," is the reply, "but we ought now to be in a condition to ask of what nature is the distinction, instead of taking it for granted as a distinction somehow lying in the existence of the facts themselves" (EW5:102–4). A stimulus, he argued, must be actively constituted by what he would later call inquiry. Consider, he says, a child reaching for a bright light who has in the past through that action obtained sometimes a toy, sometimes a piece of food, and sometimes a burn. "The question of whether to reach or abstain from reaching is the question what sort of a bright light have we here? Is it the one which means playing with one's hands, eating milk, or burning one's fingers?" (EW5:106). Far from being a stimulus-response situation, the process is one long evolving encounter between the human being and the environment, guided by inquiry and resulting in the transformation and enrichment of experience.

Written three decades later, *The Quest for Certainty* views scientific knowledge along similar lines. No matter how apparently abstract, scientific inquiry is rooted in the world. "The astronomer, chemist, botanist, start from the material of gross unanalyzed experience, that of the 'common-sense' world in which we live, suffer, act, and enjoy; from familiar stars, suns, moons, from acids, salts and metals, trees, mosses and growing plants" (LW4:138). Interpretation of this material leads inevitably to problems and questions, and to the inquiry intended to resolve them. At the same time, Dewey recognized three important differences between the simple illustration of inquiry portrayed in the previous paragraph and scientific inquiry.

First, though most human beings tend to dislike problematic situations, seeing them as obstacles to getting on with other things, scientists thrive on them. "The scientific attitude," Dewey wrote, "may almost be defined as that which is capable of enjoying the doubtful; scientific method is, in one aspect, a technique for making a productive use of doubt by converting it into operations of definite inquiry" (LW4:182). Indeed, scientists become impatient with theories that work too well and with data showing no trace of anomalies: "The thing that doesn't fit is the thing that's most interesting," Feynman once declared.[6] Scientists, Dewey knew, fear nothing more than boredom.

Second, the problematic situations of science involve interactions not between an individual and nature but between a community and nature; scientific inquiry has a social, public dimension that is comparatively absent from the inquiry of the child. It is not conducted by solitary individuals in isolated rooms speculating about nature, but by groups sharing pools of information, methods, and instruments. Contemporary scientific questions about cellular structure, stellar evolution, fundamental families of particles, high-temperature superconductors, attaining nuclear fusion, and so forth, were not formulated and pursued by individuals but by communities of investigators; a laboratory is an extremely social environment. One must be trained before entering such a community, and the training involves not only acquiring a body of skills and knowledge but also doubts and problems. The inquiry itself, the means by which it is pursued, and its reconstruction are all public and objective in that community. A single scientist working alone in a laboratory is thus as "solitary" as a single user of language. In addition, the inquiry is historical, for reconstructing the problematic situation reconstructs the community in turn, shifting the landscape of its problems and possibilities. Other social dimensions of inquiry emerge from the interaction between the scientific community of investigators and the larger social community.

The third difference between scientific inquiry and the simple inquiry of the child is the kinds of instrumentalities used to overcome the problematic situation. "If one were to trace the history of science far enough, one would reach a time in which the acts which dealt with a troublesome situation would be organic responses of a structural type together with a few acquired habits" (LW4:99)—a situation not unlike that of the child discussed in the "Reflex Arc" paper. As science evolved out of its infancy, researchers discovered that they could use tools and instruments to resolve problematic situations. "More economical and effective ways of acting were found— that is, operations which gave the desired kind of result with greater ease, less irrelevancy and less ambiguity, greater security" (LW4:100). In Dewey's view, these tools could be physical or conceptual; made of things or ideas. Nevertheless, the most elaborate and advanced laboratory technique remains but an extension and refinement of the original operations of simple inquiry. The end, too, is the same—the development and application of instrumentalities to transform the environment in the direction of greater economy and efficiency in producing changes in it, making possible a greater control over events.

In Dewey's view, inquiry is *technological*.[7] "Technology" derives from the Greek *techne*, or transformation of the environment for human purposes to supplement nature. Aristotle regarded *techne* as inferior to *episteme*, or theoretical knowledge, and productive sciences as inferior to both practical and theoretical sciences. Dewey regarded the successful production that issues from inquiry as having primacy over both theory and practical knowledge, each of which derives its significance from such production; skill divorced from it is tedious and repetitive, while theory separated from it is arbitrary and whimsical.[8] Successful production, as the outcome of inquiry, sustains both skill and theory. Yet successful production promotes the illusion that its fruits preceded it. This is the "philosophic fallacy" the taking of the results of inquiry as existing prior to it (LW1:389). In fact, inquiry creates rather than recalls these results.

For Dewey, experimentation is the introduction of changes in the world as a means to pursuing an inquiry. This introduction of changes is for the purpose of discerning relations with other changes; the discovery of a connection between a sequence of changes is useful for constructing more economical and efficient ways of controlling nature. To be sure, astronomers do not change the heavenly bodies, but they do alter the conditions under which they are seen by using different kinds of instruments at different times and in widely different points in space. "In physical and chemical matters closer to hand and capable of more direct manipulation, changes introduced affect the things under inquiry. Appliances and re-agents for bringing about variations in the things studied are employed. The progress of inquiry is identical with advance in the invention and construction of physical instrumentalities for producing, registering and measuring changes" (LW4:68). No fundamental distinction therefore exists between science and technology. To say, Dewey continues, that science is "applied" in technology and industry signifies only that "the same kind of intentional introduction and management of changes which takes place in the labora- tory is induced in the factory, the railway, and the power house" (LW4:68).

The relations between changes brought about in experimentation are represented by scientific laws and by "scientific objects," such as atoms, electrons, and the like. Here Dewey's account can become strained from the perspective of the practicing scientist, but I shall leave aside objections for the moment. Scientific laws and objects, in Dewey's view, do not corre- spond to intrinsic properties of nature but are tools used in the process of seeking its more efficient manipulation. "Laws are intellectual instrumen- talities" (LW4:164), he argued. As for "scientific objects," Dewey regarded them as "objects of the *thought* of reality, not disclosures of immanent properties of real substances. They are in particular the thought of reality from a particular point of view; the most highly generalized view of nature as a system of interconnected changes" (LW4:103). Scientific concepts and ideas hence are not statements about what is or has been, but about "acts to be performed" (LW4:111). Dewey thus sees the entire scientific enter- prise as geared toward the control of nature, and all of its concepts and methods but tools for that end. "Modern experimental science is an art of control" (LW4:80). Nature seems to have no hidden structures, no intrinsic character that might be describable in itself, apart from how we act upon it. "Nature as it exists at a given time is material for arts to be brought to bear upon it to reshape it, rather than already a finished work of art" (LW4:81).

The power of Dewey's philosophy arises from the way he sees human experience as a developmental process of inquiry in which human beings use intelligence and craft to reconstruct problematic situations, trans- forming and enriching the situation and themselves. By seeing experimen- tal inquiry as primarily concerned with performance of actions *in* the world rather than with confirmation of hypotheses *about* the world, he provides a basic perspective on experimental inquiry illuminating features that we shall find necessary to explore in an inquiry into experimentation, but which are neglected in the mythic view. These features include: the role of concrete physical events, the role of the social dimension, the ongoing rather than terminable nature of inquiry, and implications for philosophy.

Experimental Inquiry as Involving Concrete Physical Events. Just as Dewey's analysis of the child's reach for the candle highlighted features of the experience slighted by traditional psychology—head position, eye fixation, bodily twist—so his discussion of experimentation opens the possibility of exploring essential features slighted by the traditional philosophy of science. Every act, for instance, is conducted at a certain place; one could ask about the significance of experiments being conducted at particular laboratories, or about the architecture and site of laboratories. Every act is also conducted at a certain time; one could study the evolution of techniques and practices through history and their impact on experimentation. Acts can be executed skillfully or unskillfully; one could examine the nature of experimental skill. Moreover, experiments, like any other kind of human action, contain ambiguities and risks. Careful planning, good judgment, and worthy purpose do not necessarily mean a successful experiment. "Judging, planning, choice, no matter how thoroughly conducted, and action no matter how prudently executed, never are the sole determinants of any outcome. Alien and indifferent natural forces, unforeseeable conditions enter in and have a decisive voice" (LW4:6). While not news to working scientists, finding a place for such factors is difficult to do in the traditional account. Dewey's account of inquiry allows us to grant it a place.

Experimental Inquiry as Implicated in a Social Dimension. The act of discovery, Dewey says, is sometimes presented as the finding of something already there in its full being, like treasure hunters locating a chest of buried gold; instead, it involves an active modification of the world. "Discovery of America involved insertion of the newly touched land in a map of the globe. This insertion, moreover, was not merely additive, but transformative of a prior picture of the world as to its surfaces and their arrangements." The nature and place of Europe in the world was affected. "It was not simply states of consciousness or ideas inside the heads of men that were altered when America was actually discovered; the modification was one in the public meaning of the world in which men publicly act. . . . In some degree, every genuine discovery creates some such transformation of both the meanings and the existences of nature" (LW1:125). The discovery of a supernova in Cassiopeia in 1572 indicating that the heavens were not immutable; the discovery that the Earth was not cozily in the center of the Universe but somewhere in the vast reaches of space; the discoveries of evolution, relativity, quantum mechanics—all these profoundly altered the *meaning* of nature. Though less dramatic, discoveries of things like resonances and quarks altered the meaning of what it is to be a fundamental particle, and thus also our image of nature.

Another way experimental inquiry involves a social dimension is through competition with other social activities. The problematic situations addressed by scientific inquiry compete for resources with other kinds of problematic situations faced by society (defense, housing, education), and science itself can be a tool by which society attempts to reconstruct problematic situations (energy research, cures for cancer and AIDS). Problematic situations are thus socially negotiated, and both the scientific and the larger social community have a say in identifying and supporting them.

Finally, problematic situations involve a social dimension in that they are defined by the inquiry; sometimes anomalies are recognized as such, or "retrorecognized," only at a certain stage in the development of the inquiry.[9]

Experimental Inquiry as Ongoing Rather than Terminable. A third aspect is Dewey's vision of science not as the search for final answers but as a process of endless inquiry. Resolution of a particular problematic situation does not mean inquiry forever ceases. "The 'settlement' of a particular situation by a particular inquiry is no guarantee that *that* settled conclusion will always remain settled" (LW12:16). It is forever an "unfinished universe," in James's phrase, transformed in part by scientific inquiry itself; the inquiry likewise will be ever unfinished.

Implications of Experimental Inquiry for Philosophy. More is involved in inquiry into experimentation than the creation of a better description of a hitherto neglected phenomenon; Dewey thought it challenged the philosophical tradition. "If, accordingly, it can be shown that the actual procedures by which the most authentic and dependable knowledge is attained have completely surrendered the separation of knowing and doing; if it can be shown that overtly executed operations of interaction are requisite to obtain the knowledge called scientific, the chief fortress of the classic philosophical tradition crumbles into dust" (LW4:64). Predictably, Dewey was less interested in the destiny of the philosophical tradition *per se* than in consequences of its demise. The spectator view of knowledge, he notes, led to a view of science as a privileged, quasi-religious, even priestly activity; as the sanctuary of true knowledge. With this view gone, other types of knowledge could join scientific knowledge as equals. "In fact, the painter may know colors as well as the physicist; the poet may know stars, rain and clouds as well as the meteorologist; the statesman, educator and dramatist may know human nature as truly as the professional psychologist; the farmer may know soils and plants as truly as the botanist and minerologist" (LW4:176). The demythologization of science would then prepare the way for the "transfer of experimental method from the technical field of physical experience to the wider field of human life" (LW4:218), with profound implications for the social, political, and educational realms.

Dewey's account of inquiry thus provides us with what we might call the "disclosure space" of experimentation; the space in which inquirers and object of inquiry meet. The inquirers are a community of scientific researchers, sharing pools of information and practices. These inquirers pursue problematic situations that develop between their information and practices and objects of inquiry, and seek to reconstruct those problematic situations, and then to find new ones. Problematic situations can arise either through the normal pursuit of scientific inquiry, or be specified as such by the wider social community. Reconstruction of problematic situations leads to transformation in the scientific community and objects of inquiry.

Dewey does not provide us, however, with a satisfactory account of the invariance of entities in the space (their sameness amid varying conditions); nor of the role of interpretation in their appearance. Dewey insists, for instance, that scientific objects, ideas, methods, and conceptions are instruments themselves to be cast aside when improved methods are found, as the blunderbuss replaced the crossbow or the automobile the horse and buggy. This process may characterize low-level explorative research, but it awkwardly characterizes experimentation in general, in which the development of new procedures and instruments turn up "the same" entities in different manifestations. Electrons have appeared in radically different ways in different kinds of equipment for the past century, but one would not want to suggest that a different phenomenon was present each time.

The awkwardness would be removed by a satisfactory account of the presence or perception of scientific entities in the world, or more technically, by satisfactory accounts of invariance and interpretation, for to perceive a worldly presence involves apprehending invariance over variations in acts of interpretation. But it is not easy to derive such accounts from Dewey, who was afraid of hypostatizing the results of inquiry and taking them as existing prior to it, and who saw scientific concepts as devoid of "real" content and as but instruments for control of the real. "Concepts are thus simply memoranda of identical features on objects already perceived; they are conveniences, bunching together a variety of things scattered about in concrete experience" (LW4:132–33). Art rather than science is the realm of objects that are wholly "real" yet also made by human beings.

Dewey links his conviction at such moments to his battle against the mythic view: "The notion that the findings of science are a disclosure of the inherent properties of the ultimate real, of existence at large, is a survival of the older metaphysics" (LW4:83). The claim that "scientific objects" had real presence in the world led, in his view, to a serious difficulty—a problematic situation—among philosophers about how to describe the relation between the "real" and the "scientific" object. This controversy had been fueled most dramatically by Eddington's famous image, in his Gifford lectures of 1927 (just two years prior to the publication of *The Quest for Certainty*), of the existence of two tables before him, one "real" and the other "scientific."[10] Dewey's reconstruction of this issue involves the claim that the scientific table is not a table at all, for scientific objects of all kinds are simply representations of relationships of real things, and he mocks those who think otherwise: "The man who is disappointed and tragic because he cannot wear a loom is in reality no more ridiculous than are the persons who feel troubled because the objects of scientific conception of natural things have not the same uses and values as the things of direct experience" (LW4:109).

But researchers tend to experience scientific objects as having real presence in the world. Electrons are experienced as causally involved in an experiment; the real presence of the electrons is experienced as shaping the experimental event. Reinforcing belief in the real presence of scientific entities is the aesthetic or craft side of science. Experimentation is not a matter of turning on a device and reading dials; that is for lab exercises and classroom demonstrations. Nor is it a matter of the *knowing how* involved in operating a piece of machinery. It involves skillfully manipulating equipment to make something (an electron beam, an element of the immune system) appear, all the while struggling to ensure that it does not do so misleadingly or inadequately. Once it appears, it can be further manipulated and contemplated for its own sake under conditions without losing the feeling of the presence of the "same" thing.

Dewey is correct to claim that essences or intrinsic properties are disclosed only by means of extrinsic properties, but stresses this so much that he often winds up implying that nature is an infinitely pliable material that simply is what it is depending on the practices used to engage it. According to Dewey, nature has no structures that show themselves as inherently resisting or constraining our practices, or as more or less perfectly revealed in our practices, or as things to which we must accommodate ourselves rather than force to accommodate us. One finds little basis for the possibility of a critique of appearances, for the possibility that what appears may be showing itself incompletely or obscurely, and that an investigation to turn up other sides still unseen might change our view of the whole. Dewey concedes "that there is existence antecedent to search and discovery" (LW1:124). But he also counsels us to "drop the conception that knowledge is knowledge only when it is a disclosure and definition of the properties of fixed and antecedent reality" (LW4:83). What can it mean to say that existence precedes the act of knowing? Can the "same" existence appear now in one guise and now in another with different properties? If so, this allows for the possibility that human understanding can transcend the immediate appearance of something to an intuition of it as a phenomenon or being that shows itself with a different kind of presence differently in different contexts, incompletely disclosing itself in each. The kind of presence a thing has, how it "shows itself," then would depend on the world in which it shows itself, and any particular manifestation would also be a concealment insofar as other ways it could show itself would remain hidden. In an expanded inquiry into experimentation, this issue requires more clarification.[11] To pursue further the thought that scientific entities can be both indebted to the context or situation and yet also be experienced as transcendent requires a satisfactory account of invariance as well as interpretation.

In Deweyian terms, his work resulted in a successful production, one that was not only the outcome of a particular inquiry in a given problematic situation, but stood above the skills and tools used to create it and subordinated them to it. His problematic situation was shaped by the need to overcome various philosophical dualisms and combat transcendent logic and ideal forms, and by the fact that he kept his eye on the larger social context and the possibility of developing a method to transfer to it. Our own problematic situation is more modest, and it remains to discover how we can adapt the tools he provided to our context and whether additional tools are needed. To further our own inquiry we need a more specific account of invariance.

Edmund Husserl and Invariance

The beginnings of such an account can be found in the work of Edmund Husserl on perception. This work was inspired partly by Husserl's stay in Göttingen, which began in 1901 after an invitation secured by the German mathematician David Hilbert. The University of Göttingen was a world-renowned center of mathematical studies, hub of a movement that shaped the course of mathematics and physics for the rest of the century. In work that spanned several decades, the Göttingen school, which included members such as Bernhard Riemann, Felix Kline, Emmy Noether, Hermann Weyl, and Hilbert, extended and developed the concept of invariance in mathematics and physics, paving the way among other things for general relativity. Even in the 1960s and 1970s, major theoretical work in high-energy physics was based on group theoretical amplifications of the Göttingen-derived suggestion that the physically significant content of a physical theory is what remains invariant under space-time transformations.

During Husserl's Göttingen years, he gave shape to the philosophical approach he called phenomenology. Though phenomenology is generally elaborated as a philosophy of perception, it is of critical importance for the philosophy of science, though its significance in this regard is only now emerging.[12] This significance stems largely from Husserl's extension of the concept of invariance to alter profoundly our understanding of what it is to perceive. The following presentation of Husserl, like the previous one of Dewey, is brief and governed by the specific requirements of this inquiry, that is, to acquire a concept of invariance suitable for a philosophical account of experimentation.[13]

Husserl's phenomenological approach called for a return to firsthand original experience as the source of philosophical evidence. Philosophers, he thought, should look and discover rather than assume and deduce. This is not as easy and straightforward as it sounds. It does not, for instance, amount to empiricism, which assumes that objects consist of collections of observable properties gathered by the senses and organized by the mind in perception. Empiricism is an interpretation rather than a description of what happens in perception; most primordially, we see houses and hear sparrows and feel wood rather than collections of sense data that we subsequently interpret as such things. Were the empiricist's characterization of perception true, we would not experience a world at all but a chaotic flux of constantly changing impressions. Far from being based in experience, empiricism is based on suppositions about it. But the suppositions of empiricism are not untypical of those of other philosophical perspectives. To avoid importing such suppositions into philosophy, Husserl called for philosophers to take as their first task an examination of how objects give themselves most originally in experience. He was partly motivated by opposition to the neo-Kantianism then prevailing throughout German universities according to which the first philosophical task is to examine conditions of possibility of experience. But for Husserl, the thing itself teaches us what it is and how to perceive it; we must let it guide us. Thus Husserl's phenomenological slogan, "To the things themselves!"

A first discovery of this procedure is that the experience of an object is of a thing that remains the *same* amid a flux of changing conditions. To experience an object is to experience a thing with the potential to be experienced by other means, from other angles, in other circumstances, and so forth. Here Husserlian phenomenology begins to reveal its solidarity with the Göttingen approach, for the concept of invariance amid changing conditions is central to each. Just as Kline's Erlanger Program, for instance, conceived of mathematical objects that were not representable geometrically all at once but rather in definite and particular ways according to the particular planes onto which they were projected, so Husserlian phenomenology treated objects of perception, or *phenomena*, as not perceived "all of a piece" but according to the specific situation of the observer. The perceptual object, in Husserl's terminology, always gives itself to perception under a *profile*, which particular profile depending on the relative 'positioning' of observer and observed. To perceive an object as really present (rather than as an illusion or hallucination) is to experience it not simply as given under that profile, but also as having other possibilities of givenness; the object has the ability to give itself under other profiles if the respective positions of observer and observed are transformed.

Much philosophy of science, especially that of analytic orientation, assumes that human beings perceive objects first of all, and only subsequently come to learn the rules governing their behavior. Not so. To know that a thing is means to have some apprehension of the regularity of its profiles. One perceives just a single side or profile of an ordinary object—a desk, say—at any given time, but simultaneously has an apprehension of what that object would look like under transformations; when viewed from other angles and under other circumstances. For me to perceive the object as a desk and not an illusion, cardboard prop, or sculpture is to know or act as if I were to know that if I walk around it I shall see another side not now visible to me, and that I then no longer shall see this side, and that throughout all the changes it undergoes from my perspective it is still the "same" object. Such assumptions are not speculations or guesswork on my part, but basic structures of the perceptual act—of my perception of it *as* a desk. The desk is seen as a desk through a different profile in each appearance, but in each case to perceive it this way involves experiencing it as something for which the experiencing of other profiles is a possibility. My act of perceiving the desk or any object therefore involves more than apprehending the profile actually given to me; the act implicitly contains *anticipations* of other acts in which the object is experienced in other ways. It is possible that I will discover my original perception to have been misled, and my anticipations to have been mere assumptions, but this discovery is only made through sampling additional profiles.

Thus, the right method for philosophy, and what Husserl called *phenomenology,* is to put aside or "bracket" the ordinary assumptions of perception, or what he called the *natural attitude* in ideas and elsewhere, and then seek to describe phenomena by discovering correlations or regularities between their appearances to us and the character of the acts through which they are made to appear.

The regularity of profiles under transformations is the *invariance* of the phenomenon, and it can have the character of something intuited, as is the case with objects of ordinary perception, or it can be spelled out in mathematical equations, as we shall see is the case with scientific entities. It is important to be clear about the word *invariant;* I shall use it to refer to law of regularity of profiles, not to what exhibits the invariant itself, the phenomenon. The invariant is a property of the phenomenon but not a perceivable one; it shows up only in multiple appearances of the phenomenon. It is what allows the phenomenon to be recognized as a phenomenon as a certain kind. The invariant provides the law governing the sameness of what shows up in the different appearances.

The presence of invariants means that *horizons* exist which allow consciousness to focus on perceptual objects in experience. Horizons may be indeterminate and may not eventually be explored; nevertheless, they are required for a determinate object to be experienced, and in their own way horizons are given together with the object. Husserl illustrates the point with the example of a die, which is experienced as a definite object despite the fact that not all faces are given at once. "[T]he die leaves open a great variety of things pertaining to the unseen faces; yet it is already 'construed' in advance as a die, in particular as colored, rough, and the like, though each of these determinations always leaves further particulars open. This leaving open, prior to further determinings (which perhaps never take place), is a moment included in the given consciousness itself; it is precisely what makes up the 'horizon.'"[14]

Horizons, therefore, are possibilities that allow an object to be experienced as such. The process by which such horizons are formed Husserl calls *constitution*. Each conscious act is composed of several different forms of constitution, and is constituted in a different way from other kinds of conscious acts. A perceptual act has built into it several kinds of horizons consisting of several kinds of possibilities of perception—perceptions that we might have, for instance, if we manipulated the die in various ways, if we turned our gaze away from the die entirely, and so forth. Some acts of constitution are hierarchically founded on others; the constitution of time-consciousness, for instance, is foundational to all other acts of constitution. And different kinds of conscious acts—such as perception, memory, imagination, and fantasy—are constituted differently. And while some forms of constitution are passive, others (involving judgment) are active. Constitution is thus an elaborate interpretive process in which consciousness anticipates the appearance of its objects and experience becomes possible. The process of constitution does not *create* the object; I do not create the desk from nothing in perceiving it. The process of constitution does, however, create the possibility of experiencing the desk as a perceptual object. The process of constitution does not mean that my experiences are infallible; the object constituted can appear or not appear in the way expected, can fulfill or not fulfill the expectations raised by the horizons. Experiences are not infallible because one only sees a few profiles and may jump to a conclusion based on them. Upon the sampling of further profiles, the desk may turn out to be an illusion, prop, or work of sculpture. The essential phenomenological insight is that in whatever eventuality, there is no need to posit a Platonic ideal "desk" behind the appearances; the desk appears in and through each profile, each appearance, whenever the desk is perceived.

When something fulfills our expectations in a regular way, we experience it as a phenomenon. Henceforth, I shall use the word *phenomenon* in Husserl's technical sense, as something that shows different profiles in different contexts and yet still appears as "the same." Thus the word has nothing to do with phenomenalism or illusionism, nor does it carry the connotations of transience and ephemerality often associated with the adjective *phenomenal*, nor does it mean "pre-theory," as it can in physics, for instance. Experience alone allows us to discover what kind of phenomenon something is; whether what looks to be a desk is in fact so and not, say, a prop or a picture. To experience a phenomenon is to be aware of something characterized by its ability to show different facets to differently positioned

perceivers in one context, *and just those facets.* A phenomenon does not
have the invariance of an eternal eidos, but of something able to display
infinitely many profiles to infinitely many perceivers.[15] Its horizon is always open.

Two different kinds of transformations can cause the object to show
different profiles. The object itself can be transformed (e.g., the desk could
be moved), giving rise to a set of profiles whose invariant structure Husserl
calls the *noema.* Or the human body, thanks to its kinaestheses, can actively
explore the object (e.g., walk around the desk); the invariant of this set of
profiles is the *noesis.* Using the language of modern physicists, transformations are called *active* if transformations of the phenomena, and *passive* if
transformations of the observer. The point is that there is a symmetry
between the two kinds of transformations; the phenomenon itself is a
noetic-noematic structure. Another way of expressing this is to say that
the object and its manners of givenness are correlated. So basic is the
noetic-noematic correlation to phenomenology that at the end of his career
Husserl remarked that it "affected me so deeply that my whole subsequent
life-work has been dominated by the task of systematically elaborating on
this a priori of correlation."[16]

In ordinary perception, which Husserl also calls the "natural" or "naive"
attitude, the process of constitution is overlooked as one's interests and
concerns focus on the object. One looks to the desk as something to work
on, transport to another location, stand on, or hide under. In the natural
attitude, constitution and the existence of horizons remain implicit; it
would serve no purpose to be aware of them. In the *phenomenological attitude,* however, they do become the focus of attention. The aim of phenomenology, as Husserl conceived it, is to unfold and make explicit all the
various forms of constitution of conscious acts, all the silent structures that
generate the horizons in which phenomena appear. Phenomenology thus
takes on the exploration of assumptions ordinarily taken for granted—a
project that seems superfluous and unnecessary from the point of view of
the natural attitude. From Husserl's perspective, the ordinary practice of
science is emphatically in the natural attitude—as Lewis recognized when
he claimed that the "naïveté" of science is its strength. Husserl's account
of the difference between the natural and phenomenological attitudes, his
defense of the necessity of the latter, and the periodic references in his
writings to the importance of the kind of inquiry he is pursuing however
"trivial" or "obvious" it might seem thus exemplifies my earlier picture of
the philosopher as "the professional asker of the stupid question." Husserl's questions go against the grain of normal inquiry about what is real
in the world. Yet such inquiry ultimately has interest and value.

Experience, then, is not a matter of encountering a chaotic flux, nor a
matter of always encountering what we expect. Rather, it is a matter of
filling in, revising, and extending already constituted horizons in the explo-

ration of phenomena. It is only because we have anticipations that we can encounter a world of objects or experience an object in the first place. An object has to appear *as* something, *in* such and such a way, yet at the same time is associated with an invariant. Here we can see a connection between Husserl's program and our inquiry into experimentation. Like other objects, scientific objects appear differently in different situations; electrons, say, can appear in different laboratory situations in particle accelerators, beta ray detectors, Millikan oil-drop apparatus, and the like, as well as in static electricity, lightning, and the aurora borealis. But despite the different appearances, we still speak of them as appearances of the "same" object; there is an invariant. While the invariant of perceptual objects is apprehended intuitively, in scientific objects the invariant is mathematically represented in formulae. These formulae allow us to relate environments or specific laboratory practices to different kinds of appearances; they specify the law of regularity of the profiles.

We can now answer the question, asked in the previous chapter, where the invariant is located; for Husserl, it is found in both subject and object, knower and known. The structure of my apprehension of the desk, say, is the same as the structure of the way it shows itself to me, fulfilling my anticipations. The invariant thus names an identity of structures in both subject and object; it is the noetic-noematic correlation. Noetically, it programs my expectations of pendula; noematically, it describes the behavior of actually existing pendula. In more technical language, the invariant refers to the sameness of group theoretic structure of my anticipations of the object and of the group theoretic structure of the object showing itself. Isochrony, an invariant, is both a property of my anticipations of pendula, and of things that fulfill those anticipations.

This might seem the Kantian approach, but it is not. The Kantian approach does indeed involve an identity between the structures of knower and known, but requires that identity as posing the conditions of possibility of experience. We have already anticipated what we *can* perceive, and our experience can never exceed our anticipations. But in experimental praxis this is not the case, and Dewey allowed us a way to speak about the lack of fulfillment between anticipations and appearance; this creates a problematic situation that provokes inquiry and whose resolution involves a modification of the world; it affects "both the meanings and the existences of nature" (LW1:125).

While Husserl provides us with an account of invariance, we need for our purposes a more elaborate account of how what is invariant depends on the world in which it appears. This raises the issue of interpretation, of primary importance for the philosophy of science as it broaches the question of whether a philosophy of science can describe scientific activity as both disclosive of beings as well as culturally and historically bound.

Throughout much of his career, Husserl did not explicitly consider the question of the dependence of beings on the cultural and historical context, holding that the ego, or constituting activity, was "transcendental," or independent of the world. In 1935, however, at the age of seventy-six, he gave a series of lectures in Vienna and Prague interrogating the relevance of what he called the *life-world* for science. The life-world, the ultimate horizon of meaning for human activity, is historical and cultural, implying that forms of constitution founded on it are historical and cultural.[17] Husserl's views broach the theme (developed more fully by Heidegger), that meaning originates against a background context of historical and cultural origin. Even as Husserl recognized this indebtedness of meaning to worldly context, he retained the view that objects are experienced as having invariants that transcend specific worldly contexts.

Husserl's lectures were published in hurried and incomplete form under the title, *The Crisis of the European Sciences and Transcendental Phenomenology*.[18] Husserl observes that the title seems ludicrous, given the unquestionable successes of the sciences; they are "unimpeachable within the legitimacy of their methodic accomplishments." However, he continued, his motivations for speaking of a "crisis" of the sciences lay elsewhere, in their misunderstood nature and cultural role. After World War I, many people began to view science with feelings of disappointment and hostility, for it seemed to have nothing to say about the most pressing and fundamental issues: "It excludes in principle precisely the questions which man, given over in our unhappy times to the most portentous upheavals, finds the most burning: questions of the meaning or meaninglessness of the whole of this human existence" (6). Disappointed expectations led people to turn elsewhere in seeking knowledge of their place in the world, leading to increased "skepticism, irrationalism, and mysticism," manifested during Husserl's time most dangerously in the form of Naziism. The crisis thus has to do with a misunderstanding of science: "It concerns not the scientific character of the sciences but rather what they, or what science in general, had meant and could mean for human existence" (5).

When modern science began in the Renaissance, Husserl says, science *was* viewed as a single comprehensive inquiry into the important questions that could be asked about nature, society, human beings. But that vision slowly collapsed. The method chosen to pursue this inquiry was found to bear fruits only in the natural sciences; philosophy and the human sciences have lagged far behind. This apparent triumph of what was put forward as the scientific method, involving the idealization and mathematicization of nature, has given rise to misconceptions about the nature and scope of that method, and about the knowledge gained with it. Husserl thus phrases the important question to pursue as follows: *"What is the meaning of this mathematization of nature?"* (23).

In reply to his own question, Husserl tried to reconstruct the essential train of thought which led Galileo to mathematize nature. Our ordinary experience of the world reveals a relativity in the way objects are perceived at different times by different people. "But we do not think that, because of this, there are many worlds. Necessarily, we believe in *the* world, whose things only appear to us differently but are the same" (23). Yet the aim of science is to look for "true nature." Is there any method, Husserl imagines Galileo asking himself, of overcoming this relativity and discovering a "true nature"? What suggested itself was the use of geometry and the mathematics of space-time. Mathematics is exact, universal, and objective, and ever since the Greeks has been known to be applicable to the world. One might overcome the relativity of experiences of the ordinary world by treating everything in it as an example of a geometrical or mathematical type or relation—the book of Nature, to use Galileo's image, as written in mathematical symbols. "Galileo said to himself: Wherever such a methodology is developed, there we have also overcome the relativity of subjective interpretations which is, after all, essential to the empirically intuited world. For in this manner we attain an identical, nonrelative truth of which everyone who can understand and use this method can convince himself" (29).

The way to do that was twofold. First, Galileo treated geometry as related to the world through *limit-shapes* of things in the world (limit-shapes refers not just to shape but also to position and other magnitudes), with measurement conceived of as the infinitely perfectable process by which the limit-shape is determined. Husserl, in fact, defines geometry as about limit-shapes. "If we are interested in these ideal shapes and are consistently engaged in determining them and in constructing new ones out of those already determined, we are 'geometers'" (26). Second, Galileo showed that one could discover relations between these idealities—which are represented by formulae, the "predominant interest of the discovering scientist of nature" (47)—and thereby obtain true and previously unknown knowledge about bodies and events in the world. Galileo then extended this mathematical method directly or indirectly to all properties of nature.

Limit-shapes and formulae are effective tools in science, and become part of its background world; after they are developed, they can be simply taken over by successive generations. "Like all cultural acquisitions which arise out of human accomplishment, they remain objectively knowable and available without requiring that the formulation of their meaning be repeatedly and explicitly renewed. . . . It is similar to the way in which certain cultural objects (tongs, drills, etc.) are understood, simply 'seen,' with their specifically cultural properties, without any renewed process of making intuitive what gave such properties their true meaning" (26).

It is natural, Husserl says, that "the passionate interest of the natural scientist was concentrated on this decisive, fundamental aspect of the . . . total accomplishment, i.e., on the *formulae,* and on the technical method ('natural-scientific method,' 'method of the true knowledge of nature') of acquiring them and grounding them logically and compellingly for all"

(43). However, it is understandable that some "were misled into taking these formulae and their formula-meaning for the true being of nature itself" (44). This was the key move that ultimately brought about the crisis; "the surreptitious substitution of the mathematically substructed world of idealities for the only real world, the one that is actually given through perception, that is ever experienced and experienceable—our everyday life-world. This substitution was promptly passed on to his successors, the physicists of all the succeeding centuries" (48–49). The mathematical treatment of nature measures the world "for a well-fitting garb of ideas, that of the so-called objectively scientific truths," but we have come to mistake the garb for the thing itself. "It is through the garb of ideas that we take for true being what is actually a method" (51).

The transformation of "true" nature into a mathematical manifold, and the bypassing of the world as the origin of meaning, are the hallmarks of what Husserl called "Galilean science." Galilean science involves the deprecating of sense experience, and everything else that could not be mathematically handled, as *subjective* and *relative*—including, of course, "questions of the meaning or meaninglessness of the whole of this human existence." As mentioned, Husserl saw this as having adverse social consequences; he could have added that in the intellectual realm the deprecation of nonmathematical experience led to the deprecation of the role of experiment in science. If nature's 'true being' is mathematical and ideal, then theory most closely resembles nature while experimentation, which is grounded in concrete, changing worldly practices, only approximately and imperfectly approaches it. Already in Galilean science, we can say, lie the origins of the mythic view of experimentation.

The way out of the crisis, according to Husserl, is to reflect on this substitution, and show how the meaning of scientific limit-shapes and formulae arise out of and ultimately return to the life-world. But this is something scientists themselves—"unphilosophical experts"—are unable to do. "The mathematician, the natural scientist, at best a highly brilliant technician of the method—to which he owes the discoveries which are his only aim—is normally not at all able to carry out such reflections" (56–57). Indeed, scientists do not even experience the need to perform them. Scientists are like the operators of a machine that is extremely useful for some tasks, but the situation calls for individuals who are able to reflect on the possibility and necessity of such machines (52).

Husserl then begins his own attempt to conduct such reflections by developing the concept of the life-world. In the little time left before ill health forced him to stop working, he was unable to produce a thoroughly consistent and positive account of this concept; nevertheless, his remarks in the *Crisis* are instructive.[19] The life-world, he says, is the ultimate horizon of human meaning, pregiven to all in common in advance of scientific thought and philosophical questioning. It is presupposed by all human activity however specialized and of whatever kind. "For example, for the physicist it is the world in which he sees his measuring instruments, hears time-beats, estimates visible magnitudes, etc.—the world in which, furthermore, he knows himself to be included with all his activity and all his theoretical ideas" (121). Even scientific activity lies in the life-world akin to the way an object lies in a horizon:

> For example, Einstein uses the Michelson experiments and the corroboration of them by other researchers, with apparatus copied from Michelson's, with everything required in the way of scales of measurement, coincidences established, etc. Thére is no doubt that everything that enters in there— the persons, the apparatus, the room in the institute, etc.—can itself become a subject of investigation in the usual sense of objective inquiry, that of the positive sciences. But Einstein could make no use whatever of a theoretical psychological-psychophysical construction of the objective being of Mr. Michelson; rather, he made use of the human being who was accessible to him, as to everyone else in the prescientific world, as an object of straightforward experience, the human being whose existence, with this vitality, in these activities and creations within the common life-world, is always the presupposition for all of Einstein's objective-scientific lines of inquiry, projects, and accomplishments pertaining to Michelson's experiments (125–26).

Even though each particular item in the life-world may be thematized by a scientific investigation, the moment it does, such thematization removes it from the life-world, which is now the horizon of the inquiry. The life-world itself—the horizon of all human meaning—cannot itself be scientifically objectified.

Of interest is that Husserl implies that certain kinds of scientific entities are perceptible. While he says that the life-world, the "soil" of science, is "pre-scientific," meaning devoid of theoretical, ideal, mathematical models, he does not mean by this that the life-world is devoid of the precursors and residues of scientific activity; on the contrary, one can hardly deny that it contains objects that are the spinoffs of science, such as automobiles and microwaves as well as measuring instruments of all sorts.[20]

Husserl's vocabulary is deceptive here; he refers to these spinoffs of science as "theoretical results;" "theoretical results have the character of validities for the life-world, adding themselves as such to its own composition and belonging to it even before that as a horizon of possible accomplishments for developing science" (131). Husserl can not mean that spinoffs are theoretical models or need to be understood in terms of theoretical models, but rather that they are science-related ingredients of the life-world and that from them new empirical objects of science are constituted. Spinoffs may have been produced originally as the result of an intensely theoretical and mathematical inquiry, and may even be named in the life-world by the use of a theoretical and scientific semantics, but that does not mean that they are apprehended *through* such an inquiry: *microwave, computer, antenna* belong to the vocabulary of today's children. To put it another way, the life-world contains profiles of scientific objects that are not known by children, say, in a theoretical way but in a life-world way, albeit with a vocabulary that originates in theory. Children experience profiles of these scientific entities without understanding why they are profiles and what makes them profiles. This is also true for adults, of course; in science museums, the intellectually curious can experience, through unaided perception, profiles of quite sophisticated scientific phenomena without knowing why and how they are profiles; one must be told in an authoritative way what it is that the demonstrations exhibit profiles of (and there is still the possibility that the display is "rigged" to produce not profiles but what one is told are profiles). Science thus deposits naturally perceptible spinoffs in the life-world, where they are denominated by an already theoretical vocabulary.

Of interest, too, is the dual horizonality of objects that accompanies this ideal of the life-world. Husserl speaks of two basic kinds of horizons in the *Crisis*. First, objects of experience have an *internal* horizon, or one belonging to the object itself. The internal horizon consists of predelineations of the object that anticipate profiles of the same object I would experience were I to explore the object in various ways; it thus involves recognition of one kind of invariance (Maxwell's equations, say). But Husserl also considers objects to have an *external* horizon, or "place" among other things within the life-world (162). Husserl does not develop this concept, which remains abstract; it is like the skin of an apple, a necessary exterior or interface between thing and world. But it can be viewed as the background against which it appears as a foreground gestalt—a functional context or cultural niche, with any given object able to occupy a variety of such niches.[21] Consider the changing niches occupied by the electron in this century. Shortly after its discovery, at the Cavendish Laboratory in England, it was purely an object of scientific research; a popular toast at the Cavendish Laboratory annual dinner ran: "The electron: may it never be of use to anybody!" "That wish," physicist Abraham Pais observed, commenting on the tradition, "has not been fulfilled."[22] Not only did it become the object of many different kinds of research programs, in high-energy physics, solid state physics, chemistry, and other disciplines, but it also became the foundation for the twentieth-century electronics industry. A niche of this kind can be regarded as an environmental context within which the object makes itself recognizable to a human subject "tuned" to the environment. Insofar as human contexts within the life-world change, so does this place or horizon, and with it, the scheme of appearances of the object as occupying that place. Art works, likewise, are "fleshed out" differently in different historical and cultural contexts of the life-world; Handel's *Messiah* will be played differently for different audiences at different times on different instruments with different traditions of performance and different meanings. The same phenomenon may reveal itself through a characteristic set of profiles that is its internal horizon for that external horizon; there may also be a variety of external horizons or functional contexts for the given object. Internal and external horizons are each different kinds of ways in which a phenomenon "tracks" through history and culture.

In the *Crisis*, therefore, Husserl criticized Galilean science for its metaphysical assumption that the ontology of nature could be begotten by mathematics alone, bypassing the life-world, and sought to trace the activity of science to its origin in the life-world. Most post-Husserlian phenomenological literature—including that of Heidegger, Maurice Merleau-Ponty, and H. -G. Gadamer—stress this critical dimension of the *Crisis*, and understand and adapt Husserl's criticism as a criticism of science itself. But nowhere did Husserl call Galilean science the only interpretation of the scientific project; implicitly, he allowed for at least the possibility of a non-Galilean science rooted in the life-world and its perceptual objects. Still, he left no positive account of what a non-Galilean science might be.

Such an account would involve supplementing his view with a more fully elaborated conception of interpretation. Interpretation is needed both to develop Husserl's conception of the external horizon and to clarify the phenomenological position regarding the relation between measurement and theory. Even when criticizing the Galilean assumption that limit-shapes comprise the "true being of nature itself," Husserl accepts the concomitant assumption that measurement is indeed an infinitely perfectable process converging on a limit-shape, which needs to be questioned in a fully developed account of the relation between theory and experiment. Suppose, for instance, that the relation between theory and experiment is analogous to that between an ideal geometrical line and a line on a blackboard or piece of paper: one needs to know what kind of relationship is involved here. The geometrical line is not the limit that the actually drawn lines approach with better and better drawing utensils; what *counts as* a straight line is defined by the goals of the action. A segment a child draws with a crayon, a chalk stripe on a blackboard, a line traced out by a pencil on paper, the edge of a two-by-four, a laser beam—each of these can *count as* a straight line or not, depending on the uses to which the line is to be put. A range of variability exists between exemplars of acceptable lines, but that range is not mathematically specifiable; if one tried to specify it, the same problem of acceptability would crop up on another level. The question, "What is a straight line?" thus must be answered interpretively rather than mathematically.

Martin Heidegger and Interpretation

One finds an account of interpretation adequate for our purposes in the work of Martin Heidegger.[23] It is even harder with Heidegger than with Dewey and Husserl to jump into the middle of a work and borrow segments without distortion. The meanings of key words have been adjusted in accord with the meanings of other key words, and even the apparently simple discussion of tools and equipment in his magnum opus, *Being and Time* (1927), is but a momentary way station in the larger argument of the book.[24] My aim, however, is not fidelity to Heidegger's text but to the phenomenon, which a text serves in properly philosophical use.

Heidegger, like Dewey, treats human beings first and foremost as doers rather than thinkers, but in other respects their approaches differ radically. Dewey treats the activity of human beings as consisting first and foremost of responses to problematic situations within the environment. Heidegger, in his early years, treats the activity of human beings first and foremost as involved in disclosing a background context or world (what he calls Being) in which things can be encountered, discussed, investigated, ignored, forgotten, and so forth. Heidegger focuses on an entirely different dimension of relation to the world than does Dewey. For Heidegger, before there is any conscious human activity, before there is any creative activity or problematic situations, human beings are already alongside and related to things; indeed, to be a human being is to be this "related-to things." To simplify: while for Dewey human beings are *within* nature and while for Husserl they *perceive* nature, for Heidegger they are *disclosive of* nature.

Inquiry, then, does not involve a reaching out from a subjective interior to an objective exterior, but a development of a familiarity or understanding that we always already have. Heidegger calls this development of the understanding *interpretation* (188). But all understanding is finite, and takes place within specific cultural and historical traditions. While for Husserl philosophical reflection could be carried out by a transcendental ego apart from the world, for Heidegger all thinking is a finite thinking carried out by culturally and historically committed human beings. Interpretation is thus the development of a finite and already determined understanding.

Before further developing Heidegger's notion of interpretation it is first necessary to take an apparent detour and discuss his approach to what he saw to be the principal philosophical issue; namely, that of Being. When he wrote *Being and Time* he conceived this issue to be addressed by ontology. Traditional ontology begins with the observation that there are many different types of beings—tables, triangles, and poems all "are" in different ways—and proceeds by describing the various types. It answers the Being-question with something like a catalogue of the different kinds of beings and discussion of similarities and differences between them. Such a view meshes comfortably with the theological picture of the world as brought into being by a Divine author according to a blueprint that specifies the kinds of things created and their properties.

But for Heidegger, the question of Being should be posed more fundamentally. *What* a thing is, how it appears to us as a thing, depends on the context in which all things are—similar to the way, for instance, that a sound becomes a word, and hence meaningful, only within the context of a language. Husserl's inquiry, which sought critical self-consciousness about all the implicit structures of the naive attitude, was according to Heidegger still not radical enough, for it failed to address the role of the background context, and how every being is "fleshed out," so to speak, in a particular way thanks to this context, a context that provides things not just with knowability but also with reality.[25] This context in turn (again like language) depends on what is contextualized. Traditional ontology is thus incomplete without a study of this interdependence of Being and beings, what is contextualized and the context. *"Basically, all ontology, no matter how rich and compacted a system of categories it has at its disposal, remains blind and perverted from its ownmost aim, if it has not first adequately clarified the meaning of Being, and conceived this clarification as its fundamental task"* (31). Heidegger saw *Being and Time* as a first step in this clarification, and named the interrelation between Being and beings the "ontological difference."[26]

The interdependence of Being and beings is related to human activity, but we must be careful about the "who" of the activity. Different communities have different world-views, and, so to speak, see things differently. Heidegger, however, intends the "who" to refer, not just to individuals or communities, but to human existence itself. For him, human existence on the most fundamental level is already a context- or world-disclosing activity. In existing, human subjects project, inhabit, and belong to such worlds, and hence encounter objects within them. Things are not *merely* present to human beings but are *made* present through the context or world; various dimensions of this context- or world-disclosure are elaborated in *Being and Time*. This context or world is not something "behind" things, but the unlit background in which a variety of different beings can appear. Contexts and worlds are historical; as human activities change, so do they and the meaning of the things encountered in and through them.

In Heidegger's special terminology, human beings *disclose* the world. By disclosure, he means the activity of opening a context or horizon in which something can be seen as the thing that it is. "'Disclose' and 'disclosedness' will be used as technical terms in the passages that follow, and shall signify 'to lay open' and 'the character of having been laid open.' Thus 'to disclose' never means anything like 'to obtain indirectly by inference'" (105). Disclosure is not like smacking into an object in a dark room, but like generating the light and space in which a thing can be seen. To be in a world is to be involved in it and implicitly understanding the things in it; it is to see things, so to speak, in a particular kind of light.

Let me elaborate using a wonderfully simple illustration adapted from Magda King.[27] Suppose a visitor to a foreign land comes upon a strange building and, asking what it is, receives the reply, "It's a theatre." If the visitor shares the same cultural horizon as the inhabitants of the country, that answer has allowed the visitor to understand what the building *is*. But suppose that the visitor is from a different culture and asks what a theatre is. The reply may come that a theatre is a place for performing dramatic works and other spectacles. Here, too, it is possible that the visitor shares enough of the cultural horizon of the native inhabitants that this answer suffices, and the visitor has now come to understand the building as the thing it is. Then again, it is possible that the visitor does not and that future explanations are required. What form could they take? They would have to involve the elaboration of the activities of the individuals in the foreign culture. Such explanations will meet ultimate success only if they can be made to share, on some level, a context or horizon that the visitor shares with the inhabitants. Only if such a context exists will the visitor be able to recognize what the theatre *is;* a theatre is a thing only in the human world. It is not simply culture that forms this horizon, for one can come to understand and live in other cultures, but the broadest context of human existence itself. Being, therefore, is the widest possible horizon in which such understanding moves. Within this horizon, we recognize what things are—theatres, art works, churches, and so forth—together with their meaning. Moreover, this horizon can change, along with the meaning of the things that appear in it. It could be shown, for instance, that theatre has a different meaning in Greek and medieval times, in the Renaissance, and in the modern world. Scientific concepts, too, can change their meanings depending upon the background context, as for instance Kuhn has shown in the cases of "cause" and "motion."[28]

So easy is it to overlook the background context that, Heidegger says, it has been forgotten in traditional ontology. This forgetting is as natural as forgetting about the light with which we see things in favor of the things seen. But when the context is forgotten, it becomes possible to think of beings as natural kinds existing preformed over and against us. The categories of *our* world now seem to be categories of *the* world. Propositional statements become the locus of knowledge about the world, and truth or falsity a question of whether or not we have accurately represented it— whether there is a correspondence between our representations and the world. Heidegger's insight, however, is that the ability to make propositional statements capable of being true or false requires in the first place the more primordial activity by which the world has been shaped.[29] Truth in the primordial sense is disclosure. Only after disclosure has occurred can individual statements be judged true or false.

Note that both Dewey and Heidegger rebel against those who see cognition as the principal human activity, but that each sees a different origin for the urge to do so. For Dewey, the desire to make cognition the principal human activity originates in a quest for certainty held over from humanity's infancy—the psychological desire to secure something stable and permanent above the vicissitudes of existence. For Heidegger, human existence carries with it the innate temptation to forget about the context that makes possible the encounter with things, just as speakers of a language are perpetually exposed to the temptation to forget the background linguistic context that gives individual words their meaning. The forgetting of the context is an inherent temptation of the disclosure effected by the context. Whereas Dewey is concerned principally with developing an account of human inquiry in order to explore the possibilities of its transfer to different fields (ultimately, a practical aim) Heidegger is concerned in *Being and Time* with developing an account of human disclosure in order to recover the meaning of Being (an ontological aim).

This concern with the recovery of Being is apparent in sections 15–18 of *Being and Time*, in which Heidegger discusses the nature of tools and equipment. Heidegger, in fact, uses the example of tools and equipment as a kind of illustration to work out in a provisional manner themes that will occupy him for the remainder of the book. For reasons that need not concern us here, Heidegger has asked what kinds of things show themselves most immediately to human beings. His answer: tools and pieces of equipment: "In our goings-about we come across equipment for writing, sewing, working, transportation, measurement" (97). But whenever we come across such equipment, it is not truly a case of an encounter with a specific object; that piece of equipment is already familiar, experienced as belonging to an equipmental context or ensemble of tools that specifies the use of any particular tool and in which that tool plays its various roles. Furthermore, this context is not just an undifferentiated heap but consists of a complex structure in which each tool has an "assignment"; one uses a hammer to drive nails and pound sheet metal and not to knit or spread butter. The equipment context thus exists prior to any tool use and determines that use. Under what conditions is a tool there truly as the thing that it is? When, Heidegger says, the tool is used according to its assignment within the equipmental context. The hammer shows its hammerness most to a carpenter who is actually hammering. "The less we just stare at the hammer-Thing, and the more we seize hold of it and use it, the more primordial does our relationship to it become, and the more unveiledly is it encountered as that which it is—as equipment" (98). And yet, a curious thing happens. While a carpenter is hammering, neither the hammer nor the equipmental context appear to him as objects. "In such goings-about an entity of this kind is not *grasped* thematically as an occurring Thing, nor is the equipment-structure known as such even in the using" (98). This

is not to say that the carpenter is unmindful nor the hammering inatten-
tive. Far from it; "this activity is not a blind one; it has its own kind of
sight, by which our manipulation is guided" (98). Heidegger's name for
the kind of attention a user pays to tools is "circumspection" (98).

Heidegger then makes a distinction of fundamental importance not only
to his understanding of tools and equipment, but also to his concept of
interpretation and his understanding of science. Heidegger calls tools that
are manipulated appropriately within their equipmental context "ready-
to-hand." By contrast, things that are apprehended explicitly as objects
without circumspection and without any reference to a background context
are "present-at-hand." A carpenter using a hammer to make a bench, a
writer using a pen to compose an essay are grasping their *tools* ready-to-
hand; individuals who simply stare at hammers and pens without putting
them to use are apprehending *objects* present-at-hand. In the former case,
both the tool and the context within which it is used vanish from our sight
in favor of the work to be done, which, too, is ready-to-hand. "That with
which our everyday dealings proximally dwell is not the tools themselves.
On the contrary, that with which we concern ourselves primarily is the
work—that which is to be produced at the time; and this is accordingly
ready-to-hand too. The work bears with it that referential totality within
which the equipment is encountered. . . . The shoe which is to be produced is for wearing (footgear); the clock is manufactured for telling the
time. . . . A work that someone has ordered *is* only by reason of its use
and the assignment-context of entities which is discovered in using it" (99).

But Heidegger is mainly interested in the relation between the equipmental context and the appearance of the tool; he uses this relation to
model that between Being and individual beings, and to make a point
about how the former becomes visible. For how is this background context
to be seen, if it vanishes into invisibility in the grasping of things ready-
to-hand and is completely overlooked in the apprehending of things present-at-hand? How to grasp Being adequately without in the process treating it as another being?

In a transitional section (section 16) between Heidegger's discussion of
objects ready-to-hand and his discussion of objects present-at-hand, he
says that the background context shows itself, however briefly, when tools
become "conspicuous," "obtrusive," and "obstinate." These terms mark
various *stages* in the transformation from practical to theoretical understanding, from grasping a tool ready-to-hand to apprehending an object
present-at-hand. The meaning of the terms can be approximated as follows: A tool is conspicuous when our attempts to use it meet interference.
We need to call someone, but the right means is not at hand; we experience
an "un-ready-to-hand-ness." When our concern brings us to focus on what
is "un-ready-to-hand," we experience something as "un-usable;" the telephone on our desk that does not work obtrudes on us as something we

need to have work. In desperation, we may treat the telephone as something with its own workings that is in need of repair; what we are now apprehending is no longer an "un-ready-to-hand" tool, nor even an "unusable" thing, but an *ob-ject,* an obstinate and independent entity, which we are investigating in itself. Heidegger's point is to mark the stages in a highly sophisticated transformation of our understanding towards a greater distance away from the way we first, most practically, encounter it—and that while we ordinarily associate objchood with the third stage, our most direct encounter with things takes place in the first one, and that we need to recall the process of transition. But in the examination of this transition the role of the equipmental context appears. "The context of equipment is lit up, not as something never seen before, but as a totality constantly sighted beforehand in circumspection. . . . The environment announces itself afresh" (105).

Heidegger then compares the equipmental context to the understanding of the world itself. Human beings always have some familiarity or implicit understanding of the world as the context in which beings appear, which is only lit up in rare (usually problematic) situations. Indeed, the entire discussion of *Being and Time* is based on (and could not take place without) the fact that each of us already has an implicit understanding of Being.

At last we are ready to broach interpretation (sec. 32). (The philosophical discipline that investigates the general nature and varieties of interpretation is called *hermeneutics,* and the two words are often used synonymously.) In interpretation, one makes more explicit an involvement one already has and understands in some way. Let me return for a moment to the case of the conspicuous, obtrusive, and obstinate tools that interrupt the smooth execution of our projects and momentarily light up the equipmental context as something always already there. To get things back on track, we carefully examine the situation and thoughtfully plan a course of action, which we do by examining the network of assignments and functions and available resources. This making explicit of the network of involvements with which we are already involved is interpretation, and takes the form of understanding something *as* something:

> [T]he 'world' which has already been understood comes to be interpreted. The ready-to-hand comes *explicitly* into the sight which understands. All preparing, putting to rights, repairing, improving, rounding-out, are accomplished in the following way: we take apart in its "in-order-to" that which is circumspectively ready-to-hand, and we concern ourselves with it in accordance with what becomes visible through this process. That which has been circumspectively taken apart with regard to its "in-order-to", and taken apart as such—that which is *explicitly* understood—has the structure of *something as something.* . . . The 'as' makes up the structure of the explicitness of something that is understood. It constitutes the interpretation (189).

Let us recall that Heidegger uses the equipmental context as an analogue of the understanding of the world itself. The need for interpretation, let us say, arises in a problematic situation, and proceeds by developing an already-existing familiarity or implicit understanding of the world. Thus interpretation is not a matter of cloaking something with meaning, but of drawing out an already present involvement: "In interpreting, we do not, so to speak, throw a 'signification' over some naked thing which is present-at-hand, we do not stick a value on it; but when something within-the-world is encountered as such, the thing in question already has an involvement which is disclosed in our understanding of the world, and this involvement is one which gets laid out by the interpretation" (190–91). But if we inquire in order to understand, and the inquiry only makes explicit what is already known, aren't we moving in a circle? The answer, in a sense, is "yes." This is the famous hermeneutic circle, which Heidegger denies is vicious. "What is decisive is not to get out of the circle but to come into it in the right way" (195). Let me use the example of a problematic situation. Every problematic situation is permeated by understanding. Three moments of that understanding can be distinguished, which Heidegger calls forestructure of understanding: *Vorhabe, Vorsicht,* and *Vorgriff,* or forehaving, foresight, and foreconception. The *Vorhabe* or forehaving is the set of involvements that we "have" in advance without which there would be no problematic situation. Any problematic situation is also shaped by a vision or anticipation of what would reconstruct it; the spelling out of this vision is the articulation of the foreseeing or *Vorsicht* of the problematic situation. Together with forehaving and foresight is an understanding of how to initiate the reconstruction; this forms the foreconception *(Vorgriff)*. These three moments are always implicated in each other.

One might well ask how new things can be learned, given that one merely develops one's prior understanding and its forestructures. Suppose that I putter around with the telephone, unscrewing the mouthpiece, examining wires and terminals, eventually finding a loose connection that I tighten, and voila!—the telephone is fixed. Surely I have learned something, applied that knowledge, and taken a step forward. But the point is that each step of such puttering *already* is a movement of interpretation. I am not blindly messing around in such puttering; my activity has a direction, however crude. I have some notion of how to manipulate parts, I have some notion of what I have to accomplish for my puttering to stop, and I have some notion of the kinds of actions I have to do to get there (play with the wires rather than smash the telephone with a hammer, etc.). Puttering is through-and-through hermeneutical—an interpretation, a going forward, a making explicit of what I understand, to the point where eventually I may be able to use the telephone instead of cursing it. The making-explicit of this understanding in interpretation thus involves assuring, enriching, and deepening one's involvements and expectations. The

hermeneutical circle is at work in each of the three dimensions of performance in chapters four through six: in chapter four in the coming-to-appear-
ance of new phenomena in presentation (one brings new things into being
only by coming to anticipate their appearance); in chapter five in the representation of new phenomena (which is done only thanks to an anticipation
of the profiles represented); in chapter six in the recognition of new phenomena (one only recognizes the novel by being already familiar with it).

Consider, again, Galileo's encounter with the Abraham Pendulum. We
must imagine his problematic situation to consist of an uneasy sense of
something significant but concealed in the gentle sway of the lamp. His
Vorhabe included a knowledge of clocks and his pulse; his *Vorsicht* included
a knowledge of invariants; his *Vorgriff* included a knowledge to use his
pulse to measure rhythm. All three were simultaneously at work in the
way he recognized the presentation of the swinging lamp to be a profile
of an invariant that represented a property of lamps of a certain kind.

Heidegger's chief contribution to hermeneutics is in seeing it operative
not merely in textual analysis, but in our every activity. Our daily existence
involves interpretation, in which the making explicit of our understanding
leads us to discover more than we expect or even want. To be is to exist
hermeneutically. Heidegger argues that the necessity for the hermeneutic
circle is rooted in the finitude of human existence itself.

Two versions of the hermeneutical circle may be distinguished: one, *text
hermeneutics,* involves textual interpretation; the other, *act hermeneutics,* involves the performance of actions.[30] Heidegger's description of the hermeneutic circle in this section amounts to a philosophical interpretation of
our metaphor about assembling a shelf of books in order to better perceive
an object in the world. The process of assembling the shelf involves an
awareness of what we have and see already (the *Vorhabe*), what it is we are
looking to understand by adding books (the *Vorsicht*), and the kinds of
books to add to get what we are looking to understand (the *Vorgriff*). The
process is the result of a problematic situation (in our case, encountering
the problem of experimentation), and is itself interpretive.

Sections 15–18 of *Being and Time,* and the passages on interpretation,
therefore seem to provide a useful set of tools for understanding science.
In Heidegger's account of tools he outlines a set of structures that presup-
pose and make possible each other: tools, the equipmental context to
which they belong, their assignments, and the activities with which they
are used. To that list must be added the community of tool users. Heidegger presents his description as though it applied to everyone equally in
the community, but many tools—surgical implements, automobile hoists—
are so specialized as to be operable only by members of a suitably trained
community. This is especially true in science, so true that one might claim
that science does not consist primarily of a network of theories or observations, but of a set of instruments each with an assignment and belonging

to a totality, as well as a set of praxes exercised by a trained community of users. "Scientific worlds" consist of sets of practices, tools, and background knowledge. These worlds are historical in the sense that they are tied to particular times and places, and states of knowledge; the instruments, activities and world of science are always changing. Turn of the century physicists had to make their own batteries daily, used scintillation screens to detect radioactivity visually, and acquired skills like glassblowing and carpentry; in the 1930s, they employed Geiger counters and vacuum tubes and had to be skilled in electronics; in the 1980s, they use integrated circuitry and elaborate computer programs. Scientific phenomena do not appear outside of a particular scientific context or horizon, and each such horizon creates a different space in which scientific entities appear; they are "fleshed out" differently. The atomic nucleus appears differently to modern researchers equipped with particle accelerators than it did to Rutherford with his scintillation screens and Radium C. In the next chapter I shall discuss what justification one has for referring to scientific phenomena as "the same" that appear very differently in different contexts; right now the point is that such phenomena *do* appear differently, because of the different contexts to which they belong when they do appear.

Scientists may choose to explore particular possibilities opened up by the world, and through this exploration bring about changes that ultimately affect the world in turn, but they are not free to choose the field of possibilities itself; the specific scientific world into which they are born. Particularly fascinating are aporia: problematic situations, or places where the understanding of the scientific world has broken down. More specifically, aporia are disturbances in the projective activity of attempting to make sense of the world, and researchers instinctively flock to such disturbances. Aporia may begin conspicuously before becoming obtrusive and then obstinate. Consider, for instance, the saga of the discovery of parity violation in the weak interaction. It began as a result of the study of what, in the 1950s, were known as strange particles. Two particles that resulted from decays of strange particles, the tau and the theta, had the peculiarity of being identical except for a property known as parity. For various reasons, this fact complicated particle theory of the day. In 1953, many felt that a conventional way around this complication might be found. By the next year, it was realized that the problem was serious enough to threaten a breakdown of the entire theory. By 1956, physicists in search of a way out found themselves questioning hitherto sacred assumptions, and when the solution arrived, it transformed the entire theory.[31] Thus did a problem move from being conspicuous to obtrusive to obstinate; and when finally the problem was resolved it led to new concepts, activities, and revisions in the scientific "world." But even when these revisions change the scientific world so radically that one speaks of a "revolution," the process is always guided beforehand by the concepts and practices that have opened up that world in the first place.

Such an approach would also offer a way of understanding the mythic account of science. The mythic account arises, in this view, from forgetting the constitutive role of the worldly horizon in the appearance of scientific entities in our practices. During an era in which concepts and practices remain relatively stable, it is easy to forget that scientific entities appear only thanks to these historically evolved concepts and practices, and to take the historically and culturally determined present appearance for reality itself. The shared intelligibility of *this* world becomes the intelligibility of *the* world, *our* decision procedures become *the* rational decision procedures, and theory no longer addresses structures that we ourselves have created with our instruments but represents the real apart from all historical and cultural contexts. The most important part of science, in the clutch of such forgetting, thus becomes statements that express the end results of experiments on the one hand, and theories that attempt to represent the world on the other, and science's principal activity becomes that of determining the relation between these two sets of statements.

But Heidegger did not pursue his understanding of science in this manner. His approach is outlined in section 69(b) of *Being and Time*.[32] There, he rightly criticizes the approach that "understands science with regard to its results and defines it as 'something established on an interconnection of true propositions, that is, propositions counted as valid'" (408). In its place he would offer an "existential conception of science" that would treat it as an activity of human beings. For Heidegger, however, the activity of science is not one of ordinary circumspection, but of transforming what is ready-to-hand into something present-at-hand via the adoption of a "theoretical attitude," and the existential conception of science proceeds by ascertaining how this attitude originates in and arises out of circumspection. At first sight, Heidegger says, it might seem as though the theoretical attitude results from abstention from practice. Far from it; "theoretical research is not without a praxis of its own" (408). Observations with microscopes or at archaeological digs may require quite complicated preparations, and even the most abstract working out of problems requires at the very least putting pen to paper. Instead, what constitutes the theoretical attitude is a sudden changeover of the understanding of Being by which our goings-about with beings in the world has been guided, from ready-to-hand to present-at-hand (412). In the changeover, as in all conversions from ready-to-hand to present-at-hand, the object is detached from its involvement with other things ready-to-hand by overlooking its tool-character and its place so that it is now encountered as something present-at-hand. This happens, for instance, when I cease to use the hammer to pound something and consider it as an object with a mass, chemical structure, position in space-time, and so forth. But that is not all; the theoretical

attitude is not just a grasping of things as present-at-hand. In the changeover of the different sciences, different forms of the understanding of Being
are projected through a process Heidegger calls *thematization.*

Taking mathematical physics (the classic example of a modern science)
as a model, Heidegger observes that its thematization involves the projection of nature as a realm of objects with certain mathematically determinable properties such as mass. To treat the hammer as an object of physics,
I look at it as an object, but specifically with a view to its size, motion,
energy, momentum, and the like. Only after such a thematization (which,
Heidegger wants us to see, is an abstract process) has taken place can
objects of physics be measured. "Only 'in the light' of a Nature which has
been projected in this fashion can anything like a 'fact' be found and set
up for an experiment regulated and delimited in terms of this projection.
The 'grounding' of 'factual science' was possible only because the researchers understood that in principle there are no 'bare facts' " (414). The existential conception of science, Heidegger says, will arise from working out all
the details of the understanding of Being of science, and the various modes
of inquiry into the objects thus understood.[33]

Heidegger's existential conception of science envisages scientific entities
and observations as products of a prior engagement with the world; a
scientist never confronts an entity or makes an observation purely and
simply, but through a process by which that entity or observation has been
made present. For this reason, Heidegger saw science as a *derivative* activity
with respect to other activities of the life-world; the making-present involves the transformation of something ready-to-hand to something present-at-hand and thereby the detachment of scientific entities and
observations from the life-world, from human culture and history. In contrast to Dewey, Heidegger holds that the beings studied by the scientist
are desiccated.[34] "The botanist's plants are not the flowers of the hedgerow;
the 'source' which the geographer establishes for a river is not the
'springhead in the dale' " (100). Far from being world-building, science
serves to encourage eradication of experience in the world. Heidegger thus
accepts uncritically certain assumptions of the mythic account, including
the priority of theory, the detachment of the scientist from scientific entities
and observations, and the independence of scientific entities from the cultural and historical context. He does not allow, for instance, that experimental activity can disclose phenomena (make them bodily present) in the
laboratory. For Heidegger, what is disclosed by science is only theory.

These views remain largely intact in writings after *Being and Time,* with
the difference that in later works he saw the changeover as seeking to
substitute idealized, mathematicized abstractions for real objects, and is
for the sake, not of the theoretical attitude, but of turning objects into
resources for exploitation. The essence of science, Heidegger says, is "the
theory of the real." Theory projects and represents nature as a series of

eternal and abstract objects systematically related by causes expressed by laws, while "experiment is that methodology which, in its planning and execution, is supported and guided on the basis of the fundamental law laid down, in order to adduce the facts that either verify and confirm the law or deny its confirmation."[35] In such representation, Heidegger says, human beings inevitably come to conceive of natural phenomena as objects for domination and control, and "exalts himself to the posture of lord of the earth."[36] In doing science we do not seek to live *with* the world, but to achieve control *over* it. "Nature becomes a gigantic gasoline station, an energy source for modern technology and industry."[37] Through science we come to look upon nature, others, and even ourselves as calculable, manipulable resources. Beings become means to ends, and subordinating means to ends becomes the end itself. "No matter where and however deeply science investigates what-is it will never find Being."[38]

Many other phenomenologists who were either contemporaries of Heidegger or were inspired by him held similar views. Merleau-Ponty writes that science "comes face to face with the real world only at rare intervals," and that phenomenology from the first amounts to a rejection of science. "To return to things themselves," he writes, "is to return to that world which precedes knowledge, of which knowledge always *speaks*, and in relation to which every scientific schematization is an abstract and derivative sign-language, as is map-making in relation to a countryside in which we have learnt beforehand what a forest, a prairie or a river is."[39] The danger that such phenomenologists saw was that this abstraction would be substituted for the life-world and mistaken for it. Hans-Georg Gadamer writers that "modern physics has departed radically from the postulate of perceptibility that comes from our human forms of perception. . . . The impression is created that the 'world of physics' is the true world that exists in itself, the absolute object, as it were, to which all living things relate themselves, each in its own way."[40] Gadamer says that "modern theory is a tool of construction, by means of which we gather experiences together in a unified way and make it possible to dominate them."[41]

The thinkers mentioned above are only the most original and insightful of the critics of science; similar points have been made by others, including Frankfurt School members such as Max Horkheimer and Theodor Adorno.[42] The essential concern of phenomenological critics is that science develops an abstract account of the world (a map, as it were) but then forgets the origin and purpose of the map, so that it is taken for the true world and imposed back on the latter, which must be made to conform. According to Heidegger, for instance, the activity of science discloses only an abstract, theoretical description of the world (theory), and is concerned that it will be mistaken for reality itself. Science provides a blueprint for the hammer, but then neglects the practices that made it possible and efficacious to build hammers in the first place, and pretends as if the blueprint were the "real" hammer. The botanist's plant becomes the "real" plant. Heidegger wants to remind us that it is *hammering* that makes hammers, not blueprints, for if the practice of hammering substantially changes, so will the blueprints. Hence his concern to locate the origin of science in concernful activity, and his conviction that any attempt to look for an "essence" or "substance" in things is misplaced, for it implies that the being has an independence of the particular cultural and historical context in which it is found. Yet part of what science is all about is the identification of phenomena that belong to different cultural and historical contexts.

Heidegger, therefore, provides an account of interpretation, but otherwise his account of science is inadequate.

Do scientific entities appear, in experimentation, as present-at-hand objects thematized by a particular understanding of Being? No, for three reasons. First, a scientific experiment is a unique event in the world, takes time, has a beginning and end, and must be initiated and carried to completion. Every experiment is equally original, and not a recollection or echo of a past event. Nevertheless, in that unique event, something can appear able to be identified and reidentified as the same; the experiment forges the same phenomenon out of an ever-changing context, not as an externalization of something hidden or as a backwards-looking act of repetition, but as an act that is simultaneously creative and repetitive. In an experiment, something emerges, comes to light; scientific entities come to have a real presence in the world. For the experimenter in the laboratory, the "electron" is a real phenomenon, a piece of *material ontology* involved in causal explanations. And theory has something to do with that experience of sameness, in describing how this particular experience of an experimental object belongs within a particular field of possibilities of experience that are the possibilities of experience of the same object. Moreover, standardized practices can allow such phenomena to appear outside of the laboratory in the life-world where they can be used without theory, in such things as cigarette lighters, thermometers, and microwaves. Science is thus world-building, for it brings things into the life-world that were not there before—not as abstractions, but as real beings. Microwaves are not used abstractively but causally in ovens, radio waves in communication, and so forth.

Second, scientific entities are brought into the world through an *artistic* process. They are not seen with Cartesian clarity. Phenomena come to show themselves in scientific experimentation often via artistic skill. They have to be wooed, coaxed into existence, and often show up misleadingly or only inadequately; seasoned experimental scientists have troves of stories not only about how they were once almost misled into thinking that they had made great discoveries when actually something simply had been overlooked, but also about how they were misled into attributing conventional explanations to what would turn out, in the hands of others, to be genuinely new and important results. Moreover, some scientific performances are so virtuosic as to be essentially unrepeatable. Demonstration experiments such as those found in science museums—that allow the casual visitor to "repeat" famous experiments in the history of science by flipping switches—are possible only because the praxes have been standardized within a relatively nonhistorical context that is kept constant. The environment of the working laboratory, by contrast, is anything but constant.

Third, scientific entities are historically fragile. When the world changes, new scientific phenomena appear, certain existing scientific phenomena disappear, while most others show themselves in a new light. Ether, phlogiston, and caloric disappeared; will someone say that they were "really" never there at all? That would be to render incoherent the scientific world of the day. Electrons may well turn out to be manifestations of a single kind of entity called a "lepton," and that in turn may be a manifestation of something else; will some future scientist be correct to say that the electron, that fundament of twentieth century technology never "really" existed?

Scientific phenomena thus are *not* things rendered present-at-hand through a prior thematization. But neither are they things ready-to-hand, for they are thematized in some sense, explored for their own sake, and instead of becoming invisible in the praxis involving them (like the hammer) become explicitly grasped and are even lingered over.

Other thinkers of the phenomenological tradition have made additional contributions that will have to be incorporated at some point, either in the form of incorporating original ideas or corrections made to the positions expressed by the above authors. But the trio whose work I have just described address three critical elements that must be broached in an adequate account of science. These elements are: *inquiry*, or the interaction between the scientific community and environment; *invariance*, or the being of scientific phenomena as phenomena; and *interpretation* in the broadest sense, or a description of the development of finite and already determined understanding. To leave one out will render the account only partly productive.[43]

In Heidegger and Continental philosophy generally, we encounter the fear that theory, in attempting to describe invariants, is an abstraction that postulates something transhuman, eternal, and atemporal, which we are likely to mistake for being itself. This is reminiscent of Husserl's critique of Galilean science, which sought to bypass the life-world by describing reality in terms of mathematized abstractions fully describable by theory. What is needed is an account of a non-Galilean science that remains rooted in the life-world and its perceptual objects, but which also admits an adequate role for theory and invariance. The question of whether or not there is an alternative to Galilean science is of utmost importance. If science is intrinsically Galilean, as Heidegger and others think, then it is fundamentally derivative from the more primordial activity of circumspective concern. Theory has no ontological significance beyond computational power. Science is *merely successful* in its manipulation of the world, does not disclose new phenomena, and is not world-building; it is like geography, as Merleau-Ponty said, in relation to real-life forests, rivers, and prairies. If, on the other hand,, science is *not* intrinsically Galilean, then it may well bring worldly objects into the life-world, add to worldly horizons, and be world-building.

No one of the three major figures discussed above has all of the pieces for a satisfactory account of experimentation; nevertheless, they each possess elements that will be critical for that account. We may not refer *explicitly* to these authors much from now on. Nevertheless, they have provided important tools that will prove indispensable; they provide much of the *Vorhabe* for our interpretation. Dewey provided the disclosure space, Husserl the notion of what is disclosed in that space, and Heidegger the notion of how what appears in it is understood. The perspectives provided by these authors need to be supplemented; we still experience the presence of something needing to be understood, hence, a *Vorsicht*. The Deweyan perspective needs an appreciation of the experience of the actual presence of scientific entities in experimentation, and a greater sense of invariance. The Husserlian perspective needs to be supplemented with an account of how perceiving an object, scientific or otherwise, within its external horizon is a process of fulfillment or interpretation. The Heideggerian perspective needs to acquire a positive account of science that sees its activity as disclosing more than an abstract, theoretical description but new phenomena, and that does not see science as derivative with respect to other kinds of activities—that sees the aim of science as world-building rather than world-denying. And we have some rudimentary notions for how to fill in what we are looking to understand, or a *Vorgriff.*

To read Continental thinkers such as Heidegger, Husserl, and Merleau-Ponty on science may remind some of the episode in the *Pickwick Papers* in which Pickwick visits the eminent Mr. Pott of the Eatanswill Gazette, and the latter mentions an article he recently published, that had been researched in the *Encyclopaedia Britannica,* about Chinese metaphysics. Pickwick is astonished. I wasn't aware, he says, that the *Britannica* contained any articles on Chinese metaphysics. Oh, it doesn't, replies Pott. The author read for metaphysics under the letter M, for China under the letter C, and combined the information. But the arbitrariness of forcing Continental authors to address the nature of science is only apparent;

indeed, doing so is essential, for what is at stake is not only the creation of a satisfactory account of science, but also the future of Continentally-inspired philosophy. Only if Continental philosophy demonstrates that it contains the resources to develop an adequate account of science—the most powerful intellectual force of the century—will it show that it has the character of a universal rather than merely critical philosophy.

III

EXPERIMENTATION AS A
PERFORMING ART
THE THEATRICAL ANALOGY

At the end of the last century, the development of the periodic table allowed scientists to consolidate and coordinate known information about the chemical elements. Early versions were crude by modern standards. Nevertheless, merely by putting on display a certain structure among the properties of the known elements, they provided a tremendous service by exhibiting gaps and inconsistencies in existing knowledge. But periodic tables also supplied researchers with important clues regarding what research techniques could be used to fill the gaps and resolve the inconsistencies. This led, for instance, to the discovery of several new elements, such as the discovery by the Curies of polonium and radium with the aid of homologous chemistry, which involves techniques that take advantage of the fact that elements in each column of the periodic table share a similar chemical behavior.

To use an image to show how I am about to use an image, I mean to employ the theatrical analogy similarly, to consolidate and coordinate what is known (philosophically) about experimentation. In this chapter, I shall (1) discuss the argumentative use of analogy. I shall then (2) present an argument that scientific objects are perceptual objects, and explain that this entails the "primacy of the phenomenon." This will allow me, (3) to utilize the theatrical analogy to coordinate what is known about experimentation, to disclose gaps in that knowledge, and to provide clues about what kind of research techniques are needed to fill in those gaps.

Analogy

When one uses an analogy philosophically or argumentatively, one makes a point-by-point comparison between one thing and another. I shall preface discussion of the argumentative use of analogy, however, with a discussion of the closely related topic of metaphor, a figure of speech by which one thing is spoken of as if it were the other.

For Aristotle, metaphor was "the application of an alien name by transference, either from genus to species or from species to genus, or from species to species, or by analogy."[1] This definition is suspect because it assumes that all metaphors are names, that there exists a clear literal sense embodied in the name that is transferred from one area to another, and that such a transfer is indeed alien. An important and more elaborate treatment has been provided by Max Black, who calls metaphor a kind of filter.[2] To use Black's own example (drawn, in turn, from Hobbes), terminology, and analysis, "Man is a wolf" is a metaphor with two subjects, a principal subject (man) and a subsidiary subject (wolf). The metaphor conveys no meaning, of course, to those ignorant of wolves and men. However, Black points out, for the metaphor to be effective it is not necessary that individuals be experts in the study of *canis lupus* or *homo sapiens*. Far from it. The efficacy of the metaphor depends merely upon familiarity with the common background set of ideas (whether true or false does not matter) associated with the words "man" and "wolf" in the community to whom the metaphor is addressed; in fact, the metaphor will be incoherent and literally false to someone who cares only of the strict scientific properties of each species. The metaphor, "Man is a wolf," brings to bear this loose and not necessarily consistent or even true background set of ideas associated with "wolf," the secondary subject, on the background set of ideas associated with the principal subject, "man." If the metaphor is appropriate, all or nearly all of the ideas in the set associated with the secondary subject—that wolves are carnivorous, scavengers, ferocious, engaged in constant struggle, and so forth—will fit to some degree aspects of the principal subject. The metaphor Black uses, "Man is a wolf," means to be about men more than wolves.

That is not all. The metaphor's value is not that it is literally true; in fact, metaphors are literally false. That man is a wolf has been meaningfully denied.[3] But, Black says, the metaphor acts as a kind of filter with which to see afresh the principal subject, in which some aspects are highlighted, some downplayed. "Any human traits that can without undue strain be talked about in 'wolf-language' will be rendered prominent, and any that cannot will be pushed into the background." In this way, metaphors can reshape our perception of the principal subject. "The wolf-metaphor suppresses some details, emphasizes others—in short, *organizes* our view of man. . . . We can say that the principal subject is 'seen through' the metaphorical expression—or, if we prefer, that the principal subject is 'projected upon' the field of the subsidiary subject."[4] The metaphor not only causes us to see man differently, but also allows us to reinterpret what it means to be a wolf.

To say that metaphor is a kind of filter is a metaphor itself that highlights certain features and overlooks others. The filter metaphor implies that the principal subject is something complete and independent of us and our experience, and that the metaphor strains out certain features in order that others appear. What was entirely absent in this case becomes present; our recognition is now created rather than directed. Moreover, the metaphor does so not so much through words alone, but by restructuring how we see and feel something (which is why Wilshire, for instance, refers to "physiognomic metaphors").[5] The entire tissue of our experience is reoriented, and we are put in touch with aspects of that experience that were hitherto not present. Physiognomic metaphors are creative rather than filtrative. The ability of metaphors to be creative naturally depends on who hears them; for audiences with a different background set of ideas about the secondary or principal subject the metaphor obviously will have different meanings.

In a metaphor, the secondary subject is used to point out new and not otherwise distinguished or perhaps even distinguishable features in the principal subject, but without asserting an explicit deep structure. "Man is a wolf" is of this kind, for what is intended to be suggested about man is not a fully worked out set of concepts in virtue of which man is a wolf and a wolf reflects a human image but rather a loosely related collection of cues for perceiving something in a new way.

Here I mean to make a distinction between the use of metaphors and the (argumentative) use of analogies. In an argumentative use of analogy, the priority of the two terms is reversed. The secondary term, suitably adapted, is used to bring to bear a single, organized, already articulated perspective on the other term; it becomes the technically correct, albeit sometimes slightly neologic, term, and what in a metaphor held the place of the principal term is ultimately subordinated to it. Whereas two sets of background ideas are involved in the descriptive use of metaphor, only one is involved in analogies used argumentatively. Moreover, the single set of background ideas involved in an argumentative analogy already has an elaborated structure, and when the secondary subject is used to bring this structure to bear upon the primary structure, it is not in order to notice a parallel structure in a set of ideas associated with the primary subject, but to develop an appreciation of that structure in our perception of the primary subject in the first place. The argumentative analogy is thus a tool whereby a structured set of relations present in one area is introduced into another.

Argumentative analogies are a normal part of scientific thinking, and are in play whenever a set of equations is taken out of one context and adapted to another. We can take as an example the development of wave mechanics. Waves originally referred to a state of disturbance propagated from one set of particles to another—earthquakes, sound and water waves, etc. Through what amounts to an argumentative use of analogy, electromagnetic radiation came to be viewed as waves; electromagnetic radiation was the principal subject, and waves the secondary subject. The analogical (here, it is also metaphorical) calling of electromagnetic radiation "waves" enabled the bringing to bear of an entire system of knowledge, wave mechanics, on a new phenomenon, and to discern previously unnoticed aspects and relationships between aspects of the new phenomenon. The *electromagnetic radiation as wave* analogy/metaphor was thus a tool whereby the complex and interrelated set of procedures and equations comprising the existing discipline of wave mechanics was introduced into another area, optics, revealing structures of light that had hitherto been hidden. Analogies to interference, refraction, dispersion, and more were found to be present in both via a similar set of equations. Not only was the understanding of the principal subject, electromagnetic radiation, changed thereby, but also that of the secondary one, "wave." Our understanding of what a wave is changed when it was found that it could be appropriately applied to a disturbance able to be propagated in the absence of any identifiable medium whatsoever. And our understanding of wave changed still more when, via yet another argumentative analogy/metaphor, it was used in quantum mechanics, where the waves are complex functions and where the solutions of the wave equations are not themselves measurable, but are used to provide probability distributions. This kind of application of analogy underlies Bernstein's remark that "it is probably no exaggeration to say that all of theoretical physics proceeds by analogy."[6]

Argumentative analogies are also an important tool of philosophy when approaches and methods are transferred from one field of study to another. Indeed, analogy is practically a universal tool of thought, a manifestation of the Vichian principle that human beings understand the unfamiliar by the familiar.[7] The discussion of the rest of this chapter will proceed via a pair of argumentative analogies. The first involves perception as the secondary subject and the apprehension of scientific entities as the principal subject; the second involves performance as the secondary subject and the acts by which scientific entities become perceptible and present—experiments—the principal subject.

Perception and Scientific Phenomena

Shortly after a sodium cloud was discovered issuing from a volcano on Jupiter's moon Io a few years ago, an astronomer was quoted as saying that it was "the largest permanently visible feature in the solar system."[8] Robert Millikan wrote in his autobiography, about the oil-drop experiment for which he was awarded a Nobel, that "he who has seen that experiment, and hundreds of observers have observed it, [has] in effect SEEN the electron."[9] Barbara McClintock said of her research with chromosomes, "I found that the more I worked with them, the bigger and bigger [the chromosomes] got, and when I was really working with them I wasn't outside, I was down there. I was part of the system. I was right down there with them, and everything got big. I even was able to see the internal parts of the chromosomes."[10] One could cite many other examples in different fields. In 1991, British astronomers spoke of "seeing" a planet circling a pulsar when they picked up fluctuations in the radio signals emitted by the pulsar caused by gravitational effects created by the orbiting body. Recollections by colleagues of Ernest Rutherford nearly always mention his intuitive grasp of the entities he studied, which was so strong that he often spoke of "seeing" atoms, ions, electrons, and the like.

Colloquially, as these examples show, scientists often refer to their apprehension of structures of nature in the language of perception. Such language, someone will say, is only metaphorical. It reveals the existence of a highly developed intellectual faculty on the part of experimenters, an ability to grasp the presence of an object apart from its individual appearances in particular contexts with such clarity and certainty as to be analogous to the way ordinary individuals grasp the presence of perceptual objects—but what is involved is not true perception for the apprehension is mediated by instruments, inferences, and equations. To speak of perception—of seeing—is in this view a mere metaphor. The synonym for perception in each case is the secondary term, and the description of how the researchers apprehended the particular scientific entity they were studying is the principal term; the aspects underscored in the metaphor are the self-evidence, palpability, overtness, or sheer existence of the entity as it appeared to the scientists. True scientific entities, according to the mythic view, are imperceptible, for these entities are either too small to be perceived, ideal, or accessible in human experience only mediately.

In the previous chapter, Husserl's philosophical account of perception made the analogical understanding of scientific entities as perceptual objects seem possible and (at least in the *Crisis*) even plausible. It is possible, however, to argue that the application of the language of perception to knowledge of scientific objects of the sort illustrated above can be part of an argumentative analogy rather than a metaphor. Such is the thesis of Patrick Heelan, in *Space-Perception and the Philosophy of Science* and elsewhere, which puts into play a vast analogy between ordinary and scientific perception.[11] The principal intellectual debts are to Husserl, from whom Heelan borrows the basic account of perception as about invariants; to Heidegger, from whom Heelan derives the notion of the ontological and hermeneutical character of perception; and (though less explicitly) to Merleau-Ponty, from whom Heelan takes over the notion of perception as infinitely rich and never exhausted by any representation of it. Part one of *Space-Perception* is about the perception of visual space and part two about perception and the philosophy of science. In part one, Heelan argues that the shape of space in perception is not preordained but "read" like a sign system. As the signs change due to cultural and historical factors, so does our spatial perception—our "reading" of space. Whereas part one amounts to an attack on the view that ordinary space-perception is automatic and free of cultural determination, part two, "Toward a Philosophy of Science," continues the attack on the related position that perception in scientific experimentation is automatic and free of cultural determination. Heelan argues that the way we detect the presence of scientific objects in instruments, is similar to the way we read signs, signs whose meanings change over time. Just as the argument of part one is that we "read" visual cues in spatial perception, the argument of part two is that we "read" instrumental cues to perceive scientific objects. Perception in each case is a hermeneutical act, for the reading is guided by the perceived itself—by the world—rather than being completely under our control. The perceived, in fact, is infinitely richer than the language and concepts in which we talk about it, and may be used to correct that language and those concepts. This opens the possibility that we may come to see genuinely novel structures, that our world becomes enriched.

Inquiry for Heelan consists of the systematic attempt to extend perception and incorporate new perceptual structures into a pre-existing framework of understanding; inquiry is world-building. The outcome of inquiry is often what Heelan calls a "model" or "scientific model," but he distinguishes between two possible ways of interpreting the role of such models. In one, such models are understood to picture or attempt to picture the essential core of what is perceived, an essence hidden beneath the surface of the world. As we have seen, this is precisely the understanding of the role of scientific models (theories) that was criticized by Husserl, Heidegger, and Merleau-Ponty; the latter two thinkers even identified the project of constructing models thus understood as the essence of science itself. In the other way of interpreting the role of scientific models, the one adopted by Heelan, they are understood as ways of organizing sets of perceptual profiles, the profiles yielded by the reading of instrumental cues. Thus the perception and the model are linked by instrumental praxis. "This praxis . . . is the necessary but forgotten link, overlooked in the mainstream both of the philosophy of science and of analytic philosophy, between the model and the facts it represents" (21). Heelan works out the implications of this link between perception, theory, and the instrumental praxis linking them in *Space-Perception* and subsequent articles.

The praxes of science consist of techniques of preparation, measuring instruments and other scientifically disclosing equipment Heelan calls "readable technologies," and he argues that they can deliver information about the phenomenon in question just as directly as do the senses. "In 'reading' a thermometer, say, one does not proceed from a statement about the position of the mercury on a scale to infer a conclusion about the temperature of the room by a deductive argument based on thermodynamics; of course one could, but then one is not 'reading' the thermometer" (193). The reading of a thermometer to determine that the temperature is, for instance, 70°F, Heelan says, is a judgment that this is "empirical, direct, and uses scientific terms descriptively of the world." Heelan writes, "I now claim that this 'reading' is a perceptual process, since it *fulfills all the characteristics of perceptual knowledge*" (198). Heelan thus argues that scientific objects—electrons, positrons, black holes, DNA, etc.—are given to scientists "in no way different from the way common and familiar objects are given" (193) to perception in the ordinary world. Each of these announce their presence to us perceivers through readable technologies. Perception is thus generalized to all types of entities, scientific and otherwise. It thus does not matter whether the profiles of a phenomenon are mediated by instruments or not. To take a simple example (not Heelan's), lightning, static electricity, St. Elmo's fire, the aurora borealis, and shocks delivered by electric eels may all be treated as different manifestations or profiles of a natural life-world phenomenon, electricity, within a suitable theory of electromagnetism. But such a phenomenon also has other profiles that are sampled only by scientific instruments, such as by Millikan oil-drop experiments, particle accelerators, or cosmic ray detectors. The natural phenomena in this case were experienced before creation of the scientific theory that explained them as profiles of the same phenomena, but newly identified scientific phenomena can also have profiles accessible to unaided perception. Synchrotron radiation, the light given off by charged particles traveling in a charged field, is one; another is Cerenkov radiation, the phenomenon that particles moving faster than the speed of light in a medium ought to give off light. A well-known story about Cerenkov radiation has it that when British experimenter P. M. S. Blackett received the theoretical paper describing the effect, he realized that the radiation ought to be visible to the naked eye, put out the lights in his laboratory, placed some radioactive material in a test tube of water, and witnessed the characteristic blue glow. The glow was entirely accessible before the theory; it only became a profile of the scientific phenomenon in the wake of the theory.

For Heelan, then, the world of science, too, has phenomena (or, to use an older and unjustly maligned philosophical term, substances) with the same essential structure as natural phenomena; that is, they are apprehended in experience as invariants given through ordered (modelable) sets of profiles in a way constituted by the process of scientific research, theoretical and experimental. Thus the explanation for the feeling among scientists, expressed in the examples cited at the beginning of this section, that scientific entities are real things and that to apprehend them is to perceive them; the structure of their appearance in the experimental acts through which they are present is akin to the structure of the appearance of ordinary objects in ordinary perception. Experimenters can encounter things like electrons with a kind of intellectual intuition of how such objects would behave in different circumstances, a sense that an object can be grasped in some way apart from its individual appearances in particular contexts; the difference is that their encounters with the objects are mediated by instruments, meaning that a certain amount of artfulness is often involved in getting the object to appear in the first place as well as in manipulating it, and that the profiles may have to be apprehended from vastly different perspectives. Still, experimenters who have examined a particular scientific entity long enough and who have strong intuitions regarding how it would behave in different circumstances might indeed think it appropriate to say that they "perceive" it as fully as they perceive anything else. Given this approach to perception, the language used by the scientists at the beginning of this section is not metaphorical in the poetic or descriptive sense at all; rather, it is exact. One does indeed *see* electrons, chromosomes, geological epochs, sodium clouds, distant planets.

To consider scientific entities as phenomena is a much more ambitious step than even Husserl was prepared to take. Heelan treats not only "spin-offs" as perceptible objects, but *all* scientific phenomena perceptible through readable technologies. While Husserl thinks that in the constitution of the objects of science the life-world is left behind, Heelan holds that *those scientific entities are nevertheless phenomena,* though the "world" in which such phenomena appear has been enriched by the practices and instruments of a suitably trained community. These phenomena do not have to be understood in terms of theories; high temperature superconductivity, new kinds of subatomic particles, astronomical structures are often recognizable and identifiable as phenomena long before they come to be understood in terms of a mathematical model. What matters is the recognition of an invariant space-time structure. Whatever the context or external horizon, certain entities have properties that are carried with it to other contexts, even though these properties are not necessarily classical invariances such as shape or charge. A thing is something with an invariance of some kind. One would be hard put to think of how else such entities as black holes and quarks could be viewed as things. But both are apprehended in laboratory experience as invariants given by ordered sets of profiles constituted by the process of scientific research, theoretical and experimental. Thus, Heelan asserts, they may be legitimately called phenomena; colloquially, things.

This view has been challenged as unnecessary by those who emphasize the praxical or social constructivist side of science.[12] What is real, from this perspective, is simply the perceptible environmental goals that scientific practice makes achievable along with the macroscopic instruments of scientific practice—but not the postulated or "theoretical" entities that enter into the scientific representations, that are not perceptible as natural objects, and that play a role that (it is claimed) falls exclusively within that of the mathematical representation and is exclusively computational. What, then, is the necessity for the additional insistence that scientific entities themselves be perceptually apprehended as invariants or substances?

Heelan's insistence on the perceptibility of scientific entities is intended to account for why they are treated as naturalized citizens of the life-world in our environmental assessments, predictions, and explanations. Whatever may be their philosophical commitments, it is often said, scientists are realists in the lab, assuming and behaving for all the world as though they were dealing with real things with real effects. Why is this? For Heelan nothing could be more natural. Scientists treat the entities with which they deal as phenomena because they have the same structure as phenomena; that is, they are the "substantial" invariants of multiple sets of life-world appearances or profiles. If this is so, it follows that they are real in the sense of being perceptual objects. Pseudosciences such as astrology and witchcraft claim (whether successfully or not does not matter) to manage cosmic powers, not by reason of the control of phenomena in the life-world but by reason of spirits and spiritual forces that are precisely of a different kind. Science differs from such practices by treating its objects as "naturalizable" or already "naturalized" in "the world," rather like any other natural object; they have been naturalized within a suitable technologically enriched human environment. For this reason, Heelan claims, one must treat these scientific entities as having the structure of perceptual objects. Experimentation, in short, is not merely a *praxis*—an application of some skill or technique—but a *poiesis;* a bringing forth of a phenomenon.

Heelan's idea thus allows the working out of the meanings of profiles, constitution, invariance, and the rest of the phenomenological apparatus in the context of this assumption. Working out these meanings creates a schema for the philosophical investigation of experimentation utilizing the key elements mentioned in the previous chapter.

The first implication of this schema is that the criteria of objectivity for scientific entities now becomes the same as that of life-world objects. The pertinent question is: Does the object exhibit an invariance through its appearances in a sufficiently varied multiplicity of ways and under a variety of conditions? While invariances in ordinary perception are intuitions—"styles," in Merleau-Ponty's words—those of scientific objects are usually described through theories. To propose something as a scientific object is to propose a horizon of possible appearances—defined by the theory in which the object is described—and whether or not it is accepted as the phenomenon as proposed will depend on how its profiles fulfill the expectations raised by that horizon. We may find the expectations not to be fulfilled within a short period of time (the way, for instance, that the theory of relativity was not testable immediately after it was proposed), or we may be led into thinking that they are fulfilled for a while and only much later see that they are not fulfilled, as in cases of self-deception. The point is that error as well as further exploration is always revealed by the disclosure of new profiles. If I think I see the President of the United States or some other celebrity standing in the middle of the sidewalk in front of me, by stepping to the side I might obtain a profile showing me that the object was a cardboard prop instead. If it turns out that the British astronomers mentioned at the beginning of the chapter are wrong, and that the radio signals they picked up are not due to the effect of a planet on the pulsar but to rotational instability, then this will be revealed by additional profiles of the object.

In 1895, German scientist Wilhelm Röntgen announced the discovery of "X rays," and in 1903, French scientist René Blondlot announced the discover of "N rays." Each was a reputable scientist; each had made the announcement in a carefully prepared scientific paper. Both announcements raised the expectation that the respective rays would exhibit profiles in conditions of the sort that other scientists throughout the world could create in their own laboratories. That these expectations were fulfilled in the case of X Rays and were not fulfilled in the case of N rays, gave scientists cause to accept X rays as a real phenomenon and N rays as a case of self-deception. The particular sociology of the French scientific community may have held the matter in suspense for a time, but the determining factor was the lack of appearance of profiles where they were expected.[13] The same pattern is followed by more complicated cases of discovery. When the electroweak theory was proposed in the early 1970s, it implied a number of profiles in widely different areas, such as in the existence of a fourth quark, the existence of neutral weak currents, and parity violation in atoms. While the fourth quark was found in 1974, the existence of neutral currents and atomic parity violation were matters of controversy. When these controversies were ultimately settled experimentally in favor of the profiles predicted by the electroweak theory, that theory became widely accepted and the Nobel prize awarded to its creators.[14]

What is the best framework in which to consider such examples? Can we really view Röntgen's announcement, and the electroweak theory, as elaborate predictions of sense-data, and of the experiments, for instance, as attempted confirmations or falsifications of that prediction? But Röntgen and the authors of the electroweak theory had no specific idea just how the respective phenomena they had described would show up concretely in the various laboratories across the world where the claims of their theory would be examined. Is it not truer to describe these examples as cases where the existence of phenomena was proposed whose horizons indicated that profiles ought to appear in certain kinds of worldly conditions, and that when experimenters turned their attention to examining what happened in those areas where existing technologies were able to create those conditions, the appropriate kinds of profiles appeared? In the case of the electroweak theory, it happened to transpire that three kinds of readable technologies were just then being developed with a capability of creating the conditions in which the electroweak theory would exhibit profiles. These readable technologies were particle accelerators (which were just then beginning to have sufficient energy to produce a fourth quark), neutrino beams in conjunction with bubble chambers (for creating and looking at neutral currents), and dense enough beams of polarized electrons (for examining atomic parity violation). If the history of technology and instrumentation had followed a different path, other profiles revealed through other kinds of readable technologies might have become the decisive ones to sample. Does this mean that the recognition and acceptance of electroweak phenomena was haphazard, or completely dependent on social factors? Not at all. The three *positionings* were sufficient to have a good look at the way the phenomenon—electroweak unification—appeared under three widely different conditions, and the profiles exhibited a lawful behavior or invariance of the sort described by the theory. This is not to say that the theory may not have to be revised in the future, as more profiles are sampled that may deviate from the original description. In Husserl's terminology, the *noema* or invariant structure of the perceived phenomenon may have to be adjusted in the light of further examinations of the phenomenon itself, or what is apprehended as exhibiting an invariance through the noema.

A second implication of this schema concerns the correlativity of the noesis and noema in scientific phenomena. Like profiles of other phenomena, those of scientific phenomena can be changed through different *positionings* of observer and observed, a process that can involve either active or passive transformations. The different profiles that electrons reveal in cathode-ray tubes, meteorological studies of lightning, Millikan oil-drop equipment, particle accelerators, and measurements of static electricity might be considered examples of passive transformations, or transformations of the observer. The different profiles that emerge when an experimenter twiddles the dials on different runs of the same particle accelerator, changing the behavior of the beam, might be considered examples of active transformations, or transformations of the observed. As in instances of ordinary perception, the contributions of active and passive transformations are often not readily distinguishable. Moreover, scientific phenomena, too, have two different kinds of horizons, internal and external. The internal horizon consists of the profiles generated by the active and passive transformations supposing a fixed external horizon. As the horizon "belonging to the object," the internal horizon discloses the structure of the object in one of its specific historical and cultural contexts; that is, in one of its external horizons. The internal horizon thus describes the way the phenomenon is "fleshed out" in the particular worldly context in which it appears. Theory, as the description of the internal horizon, lays out under what conditions electrons appear in the contemporary context of instruments and methods. The external horizon, on the other hand, is the *place* or niche of the object among other things within the life-world amid a concrete socio-historical-technological context. It includes, for instance, a description of what counts as "fulfilling" a profile. In a certain equipmental context, a particular measurement may fall into a certain realm of acceptability and count as indicating the fulfillment of a certain profile of an electron. In a different context, a larger or smaller range of numbers will count as acceptable. But there is always the possibility that phenomena can present themselves in totally new and unexpected ways.

The internal and external horizons are thus co-dependent. Research goals may change, leading to the development of new theories, say, about electrons. Or the purpose of research might lead one to postulate that the electron has hitherto unknown properties, which may lead to its having a different external horizon. Understanding the co-dependency of internal and external horizons is crucial for appreciating the possibilities of the schema of hermeneutical phenomenology. Heidegger was afraid that any attempt to define an internal horizon was abstractive, an attempt to do traditional ontology and describe the essence of the object apart from Being, or its contextualization in a world. In the schema of hermeneutical phenomenology, however, the internal horizon is only what Husserl would have called *formal ontology*, or an abstract scheme for a being. Given a particular socio-historical-technological context, that scheme becomes "fleshed out," producing what Husserl called a *regional ontology*. In such a context, one now has the ability to name profiles and describe group structures among profiles, fulfilling the ambition of ontology while acknowledging the Heideggerian insistence on the dependence of the way objects appear on the worldly context to which it belongs in the broadest sense, or Being. The schema of hermeneutical phenomenology here proposed thus combines the insight of Husserlian phenomenology that perceptual phenomena are invariants that wear different profiles depending on the specific conditions under which they are observed, with the insight of Heideggerian hermeneutics that phenomena never appear except through historically determined practices.

Someone will object that "real" electrons nonetheless exist apart from our attempts to represent them. But the character of that "reality" is precisely what is in question here. Properties do not stand outside an object as a pulp surrounds a kernel. Beings always have a "flesh" that belongs to their very being as well as to the world in which they appear. A scientific entity does not show up in a laboratory the way an airplane shows up on a radar screen, a fully formed thing out there in the world whose presence is made known to us by a representation. Nor is a scientific entity like a smaller version of the airplane, which could be a perceptible object if only

scaled up large enough. Nor, finally, is a scientific entity like some distant
and unknown object on the radar screen that when closer becomes perceptible. A scientific entity becomes perceptible only in performance, and what
can be specified as real is only a particular noetic-noematic structure. Attempts to postulate an "ultimate" form of the object are projections.

On the one hand, then, scientific phenomena are akin to Aristotle's substance, and on the other (for Aristotelian substance does not admit historical change) to a living entity that changes over time while remaining itself
the same entity. Scientific phenomena might be thought of as having the
invariance of an Aristotelian substance, yet as showing themselves differently in different experimental setups that change over time. For laboratories are themselves cultural and historical environments, with ever-
changing equipment, personnel, projects, methods, and the like. A scientific phenomenon is thus something "behind" the data that reveal its presence and "behind" the theories by which it is described, yet never
experienced apart from either, and capable of revealing itself in new and
unanticipated ways. As independent of any particular experimental setup,
scientific phenomena are best describable through theory; as never appearing except in particular experimental setup, scientific phenomena depend on the skills and practices of scientists working in particular human
cultural and historical contexts—laboratory environments.

A third implication of this schema is that two roles can be distinguished
in the practice of science, experimental and theoretical. The experimental
role consists of the craft of preparing the conditions in which phenomena
appear, while the theoretical role consists of the formation of models, usually mathematical, for describing the lawful behavior of the profiles of phenomena that appear in experimental conditions. Experimenters describe
what appears and attest *that* it has appeared in the laboratory, while theorists explain *why* and *how* an individual phenomenon has so appeared.
The experimental and theoretical roles (Heelan's terminology for the subjects who execute these roles is S_x and S_t) are quite different. The experimenter, S_x, is concerned with the *presentation* of phenomena; with whether
or not a particular phenomenon is given in a Husserlian sense through
preparation and measurement processes involving readable technologies.
The experimenter is thus engaged in a kind of phenomenological judgment, involving a transcending of senses and categories to an intuition of
being, or something akin to what Aristotle called substance. What appears
to the experimenter are data, but data that denominate the profiles of
phenomena, say, electrons. The experimenter uses the readable technologies artfully to determine whether there is a particular invariant appearing
in or through the experimental process. This judgment is normed by a
phenomenology of perception—one that understands the object as appearing through profiles. But the judgment is also hermeneutical; it depends on historical and cultural practices. The appearance of electrons,

precisely because they are perceptual objects through readable technologies, depends in part on the changing practices of the historical and technological community. But when the phenomenon in question appears, nothing stands behind the appearances in the way an ideal "form" might; the phenomenon appears in and through the profiles, or rather, it announces itself in them.

The theorist, S_t, on the other hand, is concerned with the *representation* of phenomena. The representation does not concern a scientific entity above and beyond any of its manifestations in the lab, but rather a representation of that entity in the form of a model of its profiles. Thus the proper meaning of "explanation" on this view; a theory explains a phenomenon or how a phenomenon appears, not by describing an essence above human time and history that works "behind the scenes" of the appearances, but by constructing a representation of how its characteristic profiles emerge from the processes by which it is prepared, recognized, and measured in the laboratory. Theory explains a phenomenon by putting on display, as it were, the possible ways that the phenomenon might be "fleshed out." If and when a theory of high-temperature superconductivity is developed, it will eliminate the current trial and error procedure, and systematize experimentation by describing how other profiles might be found.

Moreover, the proper way of referring to the relation between theory and phenomenon is not correspondence but fulfillment. A theory does not correspond to a scientific phenomenon; rather, the phenomenon fulfills or does not fulfill the expectations of its appearances raised by the theory. Theory provides a language that the experimenter can use for describing or recognizing or identifying the profiles. For the theorist, the semantics of that language is mathematical; for the experimenter, the semantics is descriptive and the objects described by it are not mathematical objects but phenomena.[15] The difference between a theorist's and an experimenter's use of language might be compared to that between the use of a score to denote specific intervals and the use of the score to designate a musical performance.[16] The electron appears through a profile as having such and such characteristics, and theory provides the language for identifying the characteristics. If theory identifies an eidos, it is not of a being apart from the world (the hammer as such) but of the form of the noetic-noematic structure; that is, of a particular mode of engagement of human beings with the world. The theorist analyzes a performance or set of performances carried out on an instrumental setup. But while the experimenter "reads through" an instrumental setup to encounter a phenomenon, the theorist analyzes the performances attending to their possible representation within a field of possible representations, the way a music theorist, for instance, might analyze a performance attending to intervals, rhythms, scales, transpositions, and other such mathematically representable features of the music. Theory therefore does not describe objects that inhabit an atemporal world, but forms of this one. As something that is independent of any particular experimental setup the scientific phenomenon would be best describable through theory; as something that never appears except in a particular experimental set-up, any life-world appearance of a scientific phenomenon would depend upon the skills and practices of scientists working in a particular human historical and cultural context.

The schema presented here thus avoids the Heideggerian assumption that theory is an,abstraction which substitutes for the real above and apart from human life and history, while preserving other Heideggerian motifs. Theory attempts to describe through mathematical models a law or invariance among the profiles of a phenomenon—its internal horizon. However, insofar as those profiles never appear except in particular human cultural and historical contexts—they never appear except as "fleshed out"—a scientific theory is incomplete apart from a concrete social-historical-technological context and a standardized set of experimental praxes and cannot be formulated apart from them; this entails, in addition, that theories are themselves as historical as the social-historical-technological contexts they presume. That is not to say that some sets of equations (Maxwell's equations, say) have not been widely used in widely different social, historical and technological contexts. But the understanding of how these equations are *applied* to the situations at hand—hence, their function in a *theory* of electromagnetism—does change with the context. And even Maxwell's equations contain problematic features (such as the assumption of point charges and self-energy infinities), that eventually, in a future technological context may require substantial revision of the equations. Theory cannot be viewed as picturing entities that exist apart from the life-world. The meaning of theories is intrinsically tied up with particular standardized means of producing the phenomena they address in the life-world; thus Heelan's description of the vocabulary of theory as "praxis-laden." Experiments, meanwhile, do much more than confirm or disconfirm theories; they are performances constitutive of the content of theories. The phenomenon itself, like the piece of music, is not a material object in space and time. It only appears "in performance," and the way it appears depends on what I will call in a later chapter its *production*.

Indeed, in the schema of hermeneutical phenomenology, experimentation acquires a kind of priority over theory.[17] But this corresponds to no more than the priority phenomenology, and science itself, have traditionally placed on seeing things firsthand—the ancient meaning of *theorein*. What else can "scientific method" mean if not the artful preparation and witnessing of acts controlled by theory by and for a suitably prepared audience?

Though the experimental and theoretical roles may be executed by the same person in practice, they must not be confused. The word *electron*, for instance, differs when used by experimenter and theorist. For the experimenter, "electron" is a real phenomenon, a piece of *material ontology*, which is involved in causal explanations; the real presence of electrons in the instrumental setup is causally involved in the events that take place there. For the theorist, "electron" is an abstract term, part of a *formal ontology*, which is involved in nomological explanations; the theorist delivers an abstract model for the phenomenon consisting of a set of equations.[18]

The dual semantics is critical for the development of science, for in its absence theories could not be constrained by experimentation. If all data were theory-laden, experiments could not constrain theory because data from the start would be unintelligible without theory and could not therefore be used to test it; what data did appear would all be contextually legitimated. The independence of experimentation and theory exists in the form of this double semantics, with theory developing a syntax for representing performances, and experimentation describing the performances themselves using this representation.

Primacy of the Phenomenon

Maurice Merleau-Ponty referred to the "primacy of perception": what is primary in human experience is perception, which is richer than any attempt to represent or describe it, and which can serve to correct and add to those representations or descriptions. The schema of hermeneutical phenomenology leads to what Heelan calls the "primacy of the phenomenon": what is primary in scientific practice is the phenomenon, which is richer than the data by which it is presented as well as the theories by which it is represented or named.[19] Both *datum* and *theory* are crucial concepts in science, yet it would be an error to assume at the outset that either is primary. Such assumptions beg the fundamental questions of the philosophy of science, and betray an allegiance to prior metaphysical systems rather than to the desire to look at the things themselves. Such assumptions prevent our ability to understand what science is all about.

For antiempiricists, the role of data in knowledge of the world nearly disappears, overshadowed by the role of choosing a theory or conceptual framework. Carnap, for instance, claims that questions about the world are "internal" to one's theory or framework, and that it is meaningless to ask about the world in itself as opposed to questions about that theory or framework.[20] Quine, in one of his most famous essays, wrote that "it is misleading to speak of the empirical content [read: *data*] of an individual statement—especially if it is a statement at all remote from the experiential periphery of the field. . . . Any statement can be held true come what may, if we make drastic enough adjustments elsewhere in the system."[21]

Hilary Putnam goes further: "Objects do not exist independently of conceptual schemes. We cut up the world into objects when we introduce one or another scheme of description. Both objects and the signs [again: *data*] are internal to the description."[22] The antiempiricists are right in their emphasis on the large and important role theory plays in structuring experiments and in shaping and understanding data. But data also have a certain independence of theory; if they did not data would always exhibit the theory and the evaluation of theories through experimentation would be impossible.

How, then, does data constrain theories? Through the mediacy of the phenomenon. Data describe how a phenomenon shows itself through readable technologies; they are relative to both the way the phenomenon shows itself and to the state of the technologies. Theories, on the other hand, attempt to describe the law of behavior of the profiles that show themselves. Data, to exaggerate somewhat in order to emphasize the point against empiricists, are *fluid*. Having just one set of data of a thing is like having just a side view, a single appearance of a cup. It only comes to mean something—and to legitimate itself as a set of data, as an appearance of a phenomenon—when it is incorporated with other sets of data that present other appearances into one picture of a phenomenon as appearing through different profiles. One set of data is in the same situation as that figure, frequently used as an optical illusion, that can be seen either as a duck or rabbit profile, depending on the rest of the context; the figure only emerges as a picture *of* something, and hence as a picture at all, in the presence of that context. In the actual practice of science, data constantly have to be recalculated, reworked, or adjusted because new appearances of the phenomenon have shown additional factors in play. In practice, data often have to be thrown out because they are discovered not to have had anything to do with the appearance of the phenomenon in question. But if data are fluid, theories—again, to exaggerate somewhat in order to emphasize the point against, now, antiempiricists—are *fragile*. A phenomenon that shows itself in a way that does not fulfill the expectations of the theory invalidates the theory as a theory of that phenomenon.

Empiricists, on the other hand, tend to regard data as constituted of fixed and eternal facts; data simply emerge from the lab preformed, like cinderblocks. The primacy of the phenomenon here means that the assumption that data is an absolute foundation is misguided, like trying to account for performances on the basis of snapshots of individual moments. One needs to distinguish between three elements: the appearing of the phenomenon (in what I shall call *performance*), the phenomenon itself, and the inscription or recording of the phenomenon in performance. The phenomenon itself controls the other two. Experimentation is a physical process, and physical processes do not produce numbers (which are ideal objects) but events; data taking is the recording or inscripting of what is recognized through programmed measurement techniques. Data, like theory, is a means to get at phenomena. The meaning of data is thus relative to the quality of experimental performances giving rise to them. When scientists respond with ad hominem responses to data reports, remarking on the quality of the person or method or lab that produced them, it is not an unscientific and subjective reaction but a legitimate procedure, for it expresses awareness of the fact that all data-producing experimental performances are not equal.

Empiricism operates with the assumption that primacy belongs to observations or data, as the fundamental entities on which the work of science is based; experiment aims at data collection, theory at organizing the data collected. But the phenomenon is more than the compound of its sides— it is something which is revealed through the sides. Overemphasizing surface details, empiricism holds us too close to see the phenomenon. Antiempiricism, on the other hand, tends to attribute primacy to theoretical entities and the relations between them. But the phenomenon is more than the lawfulness of its sides—it involves an *appearance* as well. Antiempiricism holds the phenomenon too far away. The schema of hermeneutical phenomenology, however, accords to both data and theories the places where they belong; in reference to the phenomenon that they present on one hand, and represent on the other. The phenomena have primacy over both data and theories, which legitimate themselves in relation to phenomena. Data are data *for* phenomena; theories are theories *of* phenomena. Hence my somewhat exaggerated reference to the fluidity of data and the fragility of theories.

The primacy of the phenomenon also helps resolve various problems concerning the identity of scientific entities. One can speak of what is created in the experimental act in several ways. In one sense, what is created is a particular—a particular electron beam, let us say, in a particular piece of apparatus. Its identity as a particular object is perceived in the same way natural objects are perceived, with the exception that the electron beam is perceived through readable technologies. One assures oneself of its identity as a beam of electrons rather than something else through varying its profiles by changing the voltage, watching how it behaves in the presence of electric and magnetic fields, and so forth. As opposed to this identity of a particular, one can also speak of the identity of a kind, when what is created is considered as a profile of a phenomenon. This particular beam is, from this perspective, but one manifestation of the phenomenon of electrons, which can appear differently in different positionings, and in different external horizons. A phenomenon is something that has been experienced to be an invariant, which reveals itself as the same through different circumstances. Finally, one can speak of the identity of theory, or identity in connection with a particular description of the lawfulness of the profiles. The identity of the electron thus could also be described through the connectedness of its profiles in theory. The theory enunciates the structure of the noema of the electron for the community in a particular historical and technological context.

Problems arise if one takes theory rather than phenomena as primary. Traditional philosophy of science has long struggled with the problem of what has been called the *referential instability* of theories. This problem arises from the perspective of those who would make theory alone the criterion for the identity of scientific objects, not recognizing that more fundamental even than theory is the phenomenon theory seeks to represent. Theories in science are ever-changing, often with radical discontinuities. Stoney, Thomson, and Dirac each offered different concepts of the electron, and it is possible even more radically different theories will be proposed in the future. Even if many of the background concepts remain the same, and even if different theories of the same entity are sometimes possible, the character of what is described is strikingly different. One would like to say, in accordance with current scientific practice (to keep the shoe from pinching) that what the three theories describe is appropriately referred to by the same name *electrons* and that what has changed is only their appearance to us through different instrumental and theoretical contexts. But this cannot be done if one speaks of science as principally about theories conceived as representations of the world, for a different representation identifies a different object, and we cannot say that theories pick out different aspects of the same thing (as "the morning star" and "the evening star," say, may refer to the same object), for successive theories often contradict one another. One winds up in the awkward position of having to dismiss most past science as false, and of having to prepare to say the same about present-day science should future theories of a radically different character replace currently accepted ones. The implication, says Putnam, is that "just as no term used in the science of more than 50 (or whatever) years ago referred, so it will turn out that no term used now (except maybe observation terms, if there are such) refers."[23] This problem is only a special case of a general problem besetting all representationalist models of knowledge; namely, the inevitable gap between the world and the way we represent it. The gap is permanent, since it can never be guaranteed that the theory will not be revised in the future. The problem of reconciling the notion that the truth of science depends entirely on the purely cognitive reference of its ever-changing theoretical terms with the fact that science discloses scientific entities as having ever new appearances is essentially insoluble in the terms of the traditional view.

But this difficulty vanishes when one treats what is fundamental in science as *phenomena*. For at that point it becomes a matter of course that as the methods and techniques of science change, so does the appearance of phenomena and the theories we have to write for them. The schema of hermeneutical phenomenology, with its twofold horizonality, thus leads to a dynamical view of scientific entities; they are phenomena whose appearances are constituted by our praxes, whose appearances raise expectations about their other profiles, and these expectations are then refined or replaced depending on whether these profiles are fulfilled or not. Neither the changing appearances of scientific phenomena nor the changing theories by which they are described pose problems in describing their identity. In the event of a substantial shift in the outer horizon, when a phenomenon may appear markedly differently, it is still given as having had markedly different profiles in the previous outer horizon. This is part of the meaning of phenomena as historical.

Truth, in this perspective, is a disclosure of something to somebody. In this schema, that somebody is a suitably prepared audience, a community of experts prepared to see. Only after that disclosure can individual statements be judged true or false.

Another problem in traditional philosophy of science created by the lack of an adequate notion of phenomenon is that of the repeatability of experiments. In the traditional account, experiments need to be repeatable or there would be no guarantee of the truth of theories. Yet it is unclear what repetition of an experiment can mean. Each new scientific experiment differs from all others, involving different personnel, equipment, procedures, materials, computers, and so on. Even ongoing experiments often change as scientists try improvements to increase sensitivity or take advantage of new developments. How, then, can we call two different interventions into nature "the same"? This problem, too, is insoluble within the traditional perspective, because attempts to define the identity and difference of experiments are invariably done in hindsight on the outcome of those experiments—on the theories they test—rather than on the character of the experiments themselves. That is, we say that experiment Y repeats experiment X if the theory confirmed by X is also confirmed by Y; if the results differ, then either Y has uncovered a new theoretical twist or one of the two experiments is wrong.

In practice, scientists do refer to repeated experiments. One can explain this as follows: if scientific experiments constitute their objects as phenomena, then one is not under any illusion that any two experiments, any more than any two acts of perception, can be identical, nor that they need be in order to have the same phenomenon appear. One can "see" the same thing from different angles and even when it shows itself entirely differently. But if one agrees on the relevant background of equipment, practices, theories, and skills, it would make sense to say that one repeats an experiment and even that one can perform so-called "on-off" (decisive) experiments.

Philosophies of science have had a way of focusing on the invariant at the expense of the role of the cultural and historical context, or vice versa. One manifestation of focusing on the invariant takes the form of looking at science as essentially about theory, which is a distortion of the scientific process. The traditional distinction philosophers of science have made between "context of justification" and "context of discovery" serves to legitimize the neglect of the cultural and historical context. Indeed, a growing body of literature critiques this separation and strives to avoid thinking in terms of the logic of justification.[24] Another possible mistake, that of the social constructivists, is to focus too much on the cultural and historical context. This is also a distortion of scientific research activity, for it dissolves discoveries into epiphenomena of that context. With the schema of hermeneutical phenomenology and its conception of the dual horizonality of scientific phenomena, one avoids both mistakes, thus addressing what in the introduction I called the antinomic character of science. The internal horizon represents what might be called imperfectly the "essence" of the phenomenon, though this is not something atemporal, above time and history, but simply the transformation law among the profiles; the structure of the numerous ways that the object can appear within a given and possible set of praxes. The outer horizon is how the object functions in the human life-world, and corresponds to the goal to be fulfilled by scientific inquiry. Each is needed for a coherent account of scientific phenomena, and is involved in a complex relation with the other.

At this point, however, we are in danger of losing our ability to hold the discussion together. Numerous dimensions of science need to be explored, some previously mentioned: making, executing, and witnessing acts; the social character, community, and skills of science; the collaborative nature of science; the relationship between science as inquiry and science as cultural practice; the role of audience, recognition, and research programs; the relation between theories and the experimental acts based on them. One must not only find a way to address all of these, but also to do so in such a way that the discussion is organized and focused on science as itself a unified phenomenon.

A way to do so is suggested in a remark by Heelan:

> [A] theory is related to a phenomenon as a musical score is related to a musical performance. Theory alone can no more witness to the authentic presence of a phenomenon than can the score alone of a piece of music witness to the authentic presence of a musical performance: there are hosts of other relevant factors in each case. The phenomenon and the musical performance are each the realization of theoretical models or schemata that depend on social, practical, technological, hermeneutical, and artistic judgments that are local, contextual, and immersed in cultural history.[25]

This remark must be corrected slightly; theory is related to *experiment* (not to a phenomenon) as musical score is to performance. Or, theory is related to a phenomenon as a musical score is to a work of music (*the* "Seventh Symphony"). With this emendation, I shall now go one step further and introduce another argumentative analogy, likening experiments to theatrical performances. It is true that theatrical language has often been pressed into service outside of its original domain, particularly by social scientists, with results that are usually less than satisfactory.[26] However, the use of suitably adapted theatrical language allows us to treat theatrical performances as praxes in which the schema of hermeneutical phenomenology is fulfilled. Theatrical performance thus can serve as a guide for applying the schema to experimentation—in a similar way that, for instance, the examples one finds in the backs of standard physics textbooks serve as guides for applying the theories they exemplify. The *theatrical analogy* allows one to see whether and how the *theory*, so to speak, of experimental activity that is provided by the schema of hermeneutical phenomenology is fulfilled in experience, and thus lays the groundwork for forging a suitable language for addressing experimental inquiry in a philosophical manner.

The Analogy between Experimentation and Performance

By the theatrical analogy, I mean the more or less point-by-point comparison between scientific experimentation and theatrical performance. Experimentation here can mean both the uncovering of new and hitherto unknown phenomena and the continued study of known (that is, already recognized) phenomena prepared in the laboratory; the analogy will work itself out slightly differently in each case. The primary term in the analogy is thus experimentation, the secondary term performance. I mean to use the analogy as a tool whereby the complex and more or less elaborated set of concepts used in understanding performing arts are introduced into experimentation in order to reveal structures that had hitherto been hidden or obscured by the mythical view. The single set of relations that the image of the theatre is used to bring to bear on scientific experimentation is the schema of hermeneutical phenomenology just developed. I am therefore using the theatrical analogy in a fundamentally different way from others who might have used similar metaphors descriptively, without explicitly putting a philosophical canon into play.[27]

The theatrical analogy might seem forced at first, because the tradition has for the most part contrasted science and art as forms of knowing. But the appropriateness of the analogy is suggested by the common root of both "theory" and "theatre" in the ancient Greek work *theorein* (noun form *theoria*). Originally, *theorein* meant simply seeing something for oneself as opposed to hearing about it; a *theatron* was literally a place for seeing. When Herodotus tells the story of Solon's departure on his celebrated foreign voyage, the latter is said to go for the purpose of *theoria*—viewing the world firsthand.

In contemporary usage, the word *theory* has become divorced from its original sense of first-hand seeing, and one of its current meanings is "groundless speculation," virtually the opposite of the original meaning.[28] Nevertheless, aspects of experimentation still suggest the ancient sense of *theorein* as seeing something first-hand. Scientific experiments are unique events in the world undertaken for the purpose of allowing something to be *seen*. What comes to be seen is not something unique and peculiar to that event, but something that can also be seen in similar performances in other contexts. Scientific experiments can be executed well or poorly; what is seen stands out clearly in the former case and obscurely in the latter. Scientific performances are addressed to specific communities and are responses to issues raised within those communities. But properly preparing and viewing the performances requires a detached attitude, one interested in seeing what is happening for its own sake rather than for some practical end. The outcome of the detached seeing of such performances, however, can be a deepened and enriched understanding of the world and our engagement with it.

Nevertheless, this analogy is an unfortunate one to use before a philosophical audience, for philosophers have long accused the theatre of corrupting human souls and distorting truth. This distrust of the theatre began in ancient times, in the quarrel between philosophers and poets that already Plato characterized as "ancient" and in which he himself was one of the fiercest partisans. Plato's complaint was ontological; theatre imitated rather than presented. This scornful attitude persisted through Bacon, whose "idols of the theatre" refer to systems of thought that misrepresent reality and are free inventions like "so many stage-plays." "They are worlds of illusion," Bacon wrote, "created each by its own author out of his literary imagination."[29] Traces of such scorn persist even today among contemporary scholars, who, despite a few notable exceptions, choose to ignore it as a fit subject for philosophical inquiry.[30] In this book, I will not have the opportunity to do more than hint at the injustice of this understanding of theatre; for theatre (like experimentation) is in my terms first of all presentative rather than representative and confirmatory, revelatory and disclosive rather than imitative.

Nevertheless, the theatrical analogy is useful in shedding light on aspects of experimentation that have remained hidden. It is useful not merely as an organization tool to hold together the various hermeneutical dimensions previously mentioned, but also as a "physiognomic" tool, in allowing us to reorganize our perception of science and letting us see and talk about aspects of experimentation that have been overlooked in the mythical view—to *constitute*, as it were, a new perception of science itself. Against the background of the traditional philosophy of science, the theatrical analogy is bound to seem imperfect and even incoherent as a description. But it is meant to be used argumentatively as a guide to phenomenological research into scientific experimentation, to aid in "filling in" a background. Only afterwards will there be enough background for it to seem descriptive. One searches in vain in traditional textbooks of the philosophy of science for discussions, for instance, of the significance of choosing the right laboratory, of the political process needed to shepherd a proposal through a program committee, of the presentation of one's results, of getting carried away by the theatricality of it all and deceiving oneself about one's results, of selecting a team, or of the discovery process itself (often dismissed by traditional philosophy of science as only of psychological interest). Even today, few of the new approaches pose a full spectrum of questions to science, or incorporate answers to these questions in a single perspective. But in the light of this analogy, such ordinary features of science, recognizable by any practitioner, can come into the purview of the philosopher of science. The theatrical analogy will help us to appreciate the functioning within science, as in theatre, the schema of hermeneutical phenomenology.

Someone will object that science is different from theatre, that what is learned in science is different from what is learned through theatre, and that science transforms the world and individuals in it differently than theatre does. I would deny none of this. The theatrical analogy ultimately breaks down; dimensions of theatre emerge with no parallel in science, and vice versa. Theatrical performances involve embodied human beings (performers) and their actions as the focus of the performance; in experimentation, the performance is actually executed by instruments and equipment that have been set up and are operated by experimenters (who are thus not truly actors but more like producer-directors). Theatrical performances disclose, in Wilshire's words, forms of mimetic fusion—forms of the involvement of human beings with each other—while scientific performances disclose forms of involvement of phenomena in nature.[31] In science, standardization and advances in technology make possible new performance capabilities, something that has only a vague analogy in theatre. With remarkable rapidity, phenomena such a piezoelectricity, alpha scattering, and synchrotron radiation were discovered, standardized, and then turned into tools allowing performance capabilities of a much different character than ever before. Virtually any seasoned scientist has a story or two about how some procedure that in graduate school was a long and arduous process is now routinely given to technicians. Moreover, a new discovery— a genetic marker, a planetary satellite, a decay mode for a particle—is often put to use as an "anchor" or point of reference for locating and measuring other features. Other points of disconnection will turn up as well.

Someone will object that one meaning of *performance* is demonstration or show—in the sense of ostentation, drawing attention to oneself, or showing off—rather than creation. This kind of behavior is a noxious part of theatre, as in mugging or "playing to the audience," and not part of genuine scientific experimentation as opposed to instructional or public demonstrations. Or, if it is, it is minor and consists of rhetorical flourishes that one includes in talks one gives about one's results that are irrelevant to the results themselves. In *The Double Helix,* James Watson paints a wonderful scene in which Linus Pauling, in presenting his revolutionary breakthrough involving protein structure, kept his model behind a curtain throughout the talk while jumping around and "moving his arms like a magician about to pull a rabbit out of his shoe," delighting younger members of the audience and peeving senior colleagues. Pauling dramatically unveiled the model only at the very end of the talk, leading Watson to comment that it was "as if he had been in show business all his life."[32] Certainly this kind of performance—that by which results are presented to the scientific community in conferences, seminars, invited talks, etc.— has its own kind of rhetoric, its own structures, its own props (overhead transparencies as an art form), its own stars, skills, styles, and could well be studied as a branch of performance art in its own right. But demonstration or drawing attention to oneself is a derivative sense of performance with respect to its primary meaning of the skilled execution of an action in which something intended to be witnessed firsthand becomes present before a suitably prepared community.

Performances of any kind—experimental, theatrical, and otherwise— are, as I have suggested, antinomic in character. They are at once unique events *in* the world, yet at the same time reveal something *about* the world. They are at once utterly singular occurrences executed by a particular, fleeting context of materiel and personnel, and at the same time manifestations of a truth that transcends those special circumstances. To understand the antinomic character of performance, one needs what John Caputo calls a "philosophy of productivity," or account of "the creative production of unity by means of a constituting act."[33] The groundwork for such an account has been laid by the following: (1) Husserl's conception of the dual horizonality (internal and external) of perceptual phenomena, (2) the treatment of scientific entities as phenomena, via (3) the concept of readable technologies and an expanded (fully hermeneutical) conception of the external horizon, and accompanied by (4) a Deweyian-inspired notion of inquiry and the way it transforms inquirer and "world." In such a framework, scientific phenomena make their appearance perceptually through profiles in acts constituted by experimental preparations. The case of theatrical performances can be used as a guide in applying the schema of hermeneutical phenomenology to experimental activity in the course of developing a synoptic understanding of experiment.

Performances are created, witnessed, and represented as exemplars of phenomena. Phenomena can be taken as exemplars in many different ways. A hammer, for instance, may be studied as an exemplar of a tool of a certain kind, of a tool in general, of an item of merchandise, of an engineered work, of a spatio-temporal object, and so forth. One could say that to seek to take objects and events as profiles of phenomena is the *cultural attitude* of science; in the absence of this attitude there is only non- and pseudoscience. In the various branches of science this cultural attitude is worked out concretely in various ways, and in a fully elaborated inquiry into science these differences would be laid out in their similarities and differences. It is not my intention to complete such an inquiry here. I mean only to sketch out in general terms how such an attitude functions in a general way with respect to the role of performances in scientific research. To be sure, there is more to scientific culture than this particular attitude. Scientific culture includes a set of individuals, institutions, practices, products, values, goals, attitudes, educational routes, communications networks through which information is disseminated, and other socially acquired and historically changing forms of interaction of a group of people amongst each other and with nature. Many different sociological and historical inquiries can and have been made into the culture of science, with its various features and developments. From a philosophical perspective, however, what emerges as the distinguishing feature of the culture of science is the attitude that considers the experience of nature not as a succession of unique and novel surprises but of phenomena—that is, of substances which appear through different profiles in different situations—and that views nature with an eye to recognizing and representing new phenomena in it. The requirements involved in implementing this attitude—in the way of equipment, training, education, communication, and so forth—are the primary factors shaping other features of scientific culture, and as the development of science affects these requirements, scientific culture is affected in turn.

The theatrical analogy is of help here in sketching out the aspects we should inquire into, and the kinds of outcomes that would satisfy us. For theatre is also a historically changing culture with myriad aspects, centering around the activity of designing, preparing, producing, and witnessing performances of a particular kind, and is shaped by the requirements of this activity. Obvious and important differences exist, of course, between the culture of science and of theatre. The audiences to which theatrical performances are addressed include in principle everyone in the community at large; this is true even though theatre audiences actually do not consist of an even cross section of that community but largely of a subgroup of individuals whose interests, without necessarily any form of training, have led them to be particularly engaged with theatrical performances. Even a person chosen at random off the street is regarded as likely to be able to apprehend something of the significance of the events onstage. This is presuming, of course, that the person is not from a different culture; Western audiences have notorious difficulties, for instance, apprehending the significance of non-Western theatre. The audience able to apprehend the significance of experimental performances, however, is a quite specific subgroup of the community at large; that subgroup with the appropriate training. The scientific community, therefore, is a *suitably prepared* community, which has undergone a training process enabling it to recognize the significance of experimental performances. The forms of preparation vary, of course, depending on the particular branch of science; a biologist will not be suitably prepared to appreciate the experimental performances in a physics laboratory, for instance, even though the presence of a shared scientific culture may allow the biologist to appreciate more quickly than nonscientists the role of the research program, the technical and methodological strategies, the experimental difficulties, and so forth. Even within each branch of science there are numerous communities, some nested in others. As microbiologist Ludwik Fleck has described in *Genesis and Development of a Scientific Fact*, there have been cases where such suitably prepared communities able to execute and appreciate the significance of certain experimental performances have consisted of a handful of researchers within a single laboratory.[34]

One can now begin to develop the argumentative analogy between experimentation and performance to consolidate and coordinate what we already know about experimentation. Here are some of the topics on which we already have an extensive literature: theory, explanation, laws, confirmation, scientific terms, concept formation, probability, induction, deduction, social influences on science, case studies of experiments, histories of science and technology. These spots are already filled in on our "periodic" table.

Most of these topics assume the givenness of the scientific explanandum. But the explanandum is not given antecedently to the experimental process. And experimentation is such a large part of science that the absence of an account of it creates not just one but numerous "holes" in our periodic table. These holes can become visible, I have suggested, with the aid of an argumentative analogy with performance. For performance has three principal dimensions; presentation, representation, and recognition. Performance is first of all an execution of an action in the world which is a *presentation* of a phenomenon; that action is related to a *representation* (for example, a text, script, scenario, or book), using a semiotic system (such as a language, a scheme of notation, a mathematical system); finally, a performance springs from and is presented to a suitably prepared local (historically and culturally bound) community which *recognizes* new phenomena in it. The field develops through an interaction of all three. The important "holes" in our account of experimentation that need exploration thus include: (1) a (philosophical) account of the way in which phenomena are presented in experimentations, or how "data" and "evidence" appear in the first place; (2) an account of how such phenomena are represented, including the link between instrumental praxes and model; and (3) an account of how new phenomena are recognized or discovered. The accounts of inquiry, invariance, and interpretation given in the preceding chapter will help to fill in these three holes, which I shall do in the next three chapters.

But performance cannot be thought of in terms of its product alone; performances must be *prepared* by an advance set of behaviors and decisions. It is a mistake, for instance, to think of a theatrical performance as exhausted in a video made of the opening night, which could then be replayed, rewound, and replayed again at will. A spectrum of decisions and activities have to take place before that opening night in order to "shape" that opening night performance. (The large number of individuals, different kinds of skills, and amount and variety of resources involved in major theatrical productions is a principle reason why I have chosen them rather than musical performances in my analogy.) If these decisions and activities were made differently, the outcome would differ as well. It is equally a mistake to think of experimentation as exhausted in the data that come from it. Experimentation is a process that reveals data, and the process takes time and is effected in phases. It involves its own set of decisions and activities made before the first experimental run, and how these decisions and activities are carried out affect the outcome of that run. The conditions of an experiment, like that of a theatrical production, do not have one solution.

It is therefore necessary to consider *production,* another "hole" in our account; production refers to the set of decisions made in advance of a performance necessary for it to take place at all, and which makes it possible to speak of many of "the same" kind of performance. This I shall discuss in chapter seven. What is paramount, however, is the performance itself; just as there is a primacy of perception that dominates appearing and representing, and a primacy of the phenomenon that dominates data and theory, so there is a *primacy of performance* that dominates each of the dimensions just mentioned. A true performance dominates performers as well as audience. It has a reality of its own that one "lives up to," and one distorts this reality if one attempts to reduce the meaning of performance to a series of actions or gestures on the one hand, or to theory or representation on the other. This is related to the kind of primacy we saw Dewey extending to successful productions over theory and skill.

The theatrical analogy is nothing to be "proven." Proof is the point of arrival for arguments in disciplines involving ideal entities such as mathematics. But in disciplines that aim at the discovery of novel worldly structures, disclosure rather than proof is the culminating event. The theatrical analogy is a tool to create a research program about the nature of experimental science. It is an attempt to design the questions that should arise within such a process of inquiry, and anticipate the kinds of answers that would satisfy us. The theatrical analogy, like any argumentative analogy, is not a finished thing, but belongs to an inquiry, a process; its value depends on what that process is able to disclose about the nature of experimentation.

IV

PERFORMANCE
PRESENTATION

Presentation is that dimension of performance which aims at achieving the presence of a phenomenon under one of its profiles; a failed presentation is one in which this presence does not transpire. The relevant philosophical literature concerns the nature of actions in general. One would imagine this to be a theme of the philosophy of action. But the kind of action theory required here has little to do with what is usually found in the body of philosophical writings that go under that name. A theory of action suitable for understanding experimentation must explore what is involved in planning and skillfully executing acts. We must also approach action in a different way from Dewey, Mead, and other pragmatically inspired philosophers, who generally focus on the social aspects of action.[1]

One kind of action particularly important in understanding performance is play, or a back and forth motion between a player or groups of players and things that are played upon (instruments, say, or equipment and field), so that something (a piece of music, a game) is "played out." This inquiry, too, as a kind of performance, will need to engage in play, which here will take the form of a back and forth motion between speculations about scientific activity and descriptions of it. Because the speculations and descriptions are in different "voices," they are rendered in different typefaces.

Laboratories

Like acts of perception, acts of performance are constituted or prepared by both passive and active forms. *Passive* forms consist of readable technologies; reliable, standardized technologies whose operation one takes for granted and through which the world may be "read." The horizon for readable technologies is provided by the laboratory.[2]

An unsuspecting visitor to a modern-day high-energy physics experimental hall could be forgiven for taking it to be a disorderly warehouse. The

103

overhead crane, the single most prominent feature of an experimental hall proper, runs on rails spanning the entire length of the building and is made to bear the weight of heavy concrete blocks. A series of roll-up doors along one end may be present, which are occasionally opened to admit a fleet of forklifts to move material in and out. Cable and pipe trays line the walls, while embedded in the floor of the hall every dozen or so feet is a series of tunnels so that the cables and pipes can be readily conveyed to every spot within. Concrete blocks sit atop each other in odd accumulations; every so often makeshift, erector set-like stairs lead up and over them. Wild tumbles of pipes and cables cascade from the piles and then just as abruptly vanish again. Here and there stand racks of computers and power supplies, bottles of gas, portable radiation monitors whose lights are flashing (thankfully!) green, telephone booths, and other equipment. Walking through the warren of small paths that lead throughout the hall, one encounters every so often a trailer—usually perched high atop a pyramid of equipment and concrete blocks—where experimenters can gather when not working on the floor of the hall. On one such trailer appears a testimony to the apparent chaos: a makeshift sign reading, "If racoon spotted again, call target desk!" But a more experienced look reveals the hall and everything in it to have a quite specific structure. The cables and pipes ferry power, information, water, and gases from sources outside the hall to equipment inside. The concrete blocks surround a beam carrying protons from the particle accelerator itself and are intended to absorb the radioactivity that emerges from the beam. The function of the concrete blocks is further evident in their composition (iron-rich, to make them more effective radiation shields), shape (all are built to be stacked conveniently on others), and structure (many have tubes embedded in them through which cables can be threaded if desired). Each block also has a recess in which is implanted a pipe that can be affixed to a hook suspended from the overhead crane. Even the safety precautions, from shielding to radiation monitors to fire equipment, are related to the specific kinds of dangers likely to be associated with such events for the human beings who need to be present in their preparation and execution.

The hall has been constructed specifically for the purpose of facilitating the performance and witnessing of a specific kind of action therein. It is a *theatron,* or place for enacting and seeing a performance; better still, given that the stage is everywhere and the *audience* (more on this in a moment) as it were on the outskirts, an *amphi-theatron.* The kind of action involves the creation and detection of subatomic particles, and all the equipment inside has been designed to facilitate the execution of these events. To use Heidegger's terminology, everything in the hall has an assignment, and the experimental hall itself serves as an accomodating horizon to these equipmental assignments. The sets of equipment and the praxes are his-torical, for they are tied to particular times, places, and states of knowledge; while an experimental hall in the early 1950s might have emulsions, scintillators, and cloud chambers, one in the 1960s was likely to have spark chambers and bubble chambers instead. Other kinds of laboratories, too, are tailored to suit the particular kinds of performances that take place in them—even that of the telescope.

> *Optical telescopes are located in laboratories as far as possible from lights and people, preferably on a high mountain and in a desert climate free of clouds and water vapor. The pier on which the typical modern optical telescope is mounted stands anchored to bedrock by a solid mass of thousands of tons of concrete. This pier is completely isolated from the rest of the observatory, which is built on a second pier, wrapped around the first, in order to prevent vibrations in the rest of the building from being transmitted to the telescope pier. Around the telescope pier, too, is a rapid ventilation system, which prevents temperature differentials from developing between the air in and outside the building. All these characteristic features of an optical astronomical observatory correspond to the need to facilitate the particular kind of action that takes place in them. Light passes down the telescope tube, reflects through a system of mirrors (depending on the arrangement, generally four or five), and strikes an electronic detector; photons from the light beam excite electrons in a semiconducting material in the detector and cause them to move from lower energy wells to higher ones, creating a set of electrical impulses that can be analyzed by computer to localize the source of the light; this information also can be transformed into an image if desired. Even before this action is performed, however, a specially constructed environment has been prepared for that performance.*

Concomitant to its neglect of experiment, traditional philosophy of science has overlooked the significance of the laboratory. It has assumed, in effect, that the laboratory is a place where one eavesdrops, so to speak, on a *scientific world* of imperceptible entities that govern the events that take place in our own world. We become transported to a realm where we can 'check' our theories describing the structure of this more fundamental world by somehow holding them up against an exemplar. An equivalent view of theatre would have it that entering a theatre amounts to entry into a special *theatrical world* peopled by real Hamlets, Ophelias, Luckys, Pozzos and the like, represented by actors but not presented by them—a fantasy place which, though not our own world, can tell us something useful about it. But the real process is both more mundane and more extraordinary than that. The laboratory, like the theatre, is a special place where special things are learned. But those special things can be learned only because the laboratory has been specially constructed to execute and witness particular kinds of actions.

The scientific object has been *prepared* for study in this environment.

This preparation is similar to what Husserl referred to as constitution but I shall not use that word because of the idealistic connotations that it now carries and in order to emphasize that experimentation is a process that involves bringing something materially into being. While some forms of preparation are passive, others (involving judgment) are active; it's not all a matter of securing reliable equipment but can involve application of skill. The process of preparation does not create the scientific object, but does create the possibility of experiencing it as a perceptual object.

The laboratory itself is a particular *space of action* that serves as the point of departure for such preparation, which is a particular goal-oriented activity.[3] It includes various passive forms of preparation of scientific activity. Like venues for theatrical performances, different kinds of locations exist for experimental performances with different passive forms of preparation. As the differences between the high energy physics hall and the observatory show, functional differences give rise to different kinds of laboratory environments, but it is also possible to trace historical changes in the size, architecture, and organization of laboratories. Laboratories built in the last century incorporated auditoriums for the interested and educated public to witness experiments, reflecting a different cultural experience from our own; today, national laboratories are apt to have instead technology transfer offices, public relations departments, day care centers, science museums, education centers, archivists and historians, reflecting an entirely different view of the laboratory and its relation to society. As with every other institution, the presence of a laboratory in a particular location can be appropriately studied as a product of social, political, economic, and personal factors.[4] In Deweyian terms, laboratories themselves are socially negotiated outcomes of the reconstructions of problematic situations, and both the scientific community and the larger social and political communities have a say in identifying the character of the problematic situation as well as in its reconstruction.

Obvious differences exist, of course, between theatrical and scientific theatres. Let us recall that it is a question here of observing the performances of equipment rather than human beings (except, as in psychology experiments, in much different ways than are involved in theatre). Experimenters, also, are in the role of producer-directors. Moreover, perceiving something through readable technologies means that one does not have to be in the physical proximity of the event itself to witness it—indeed, events in, say, a particle detector could not be directly witnessed even if one somehow had access to the interior by window—but one must rather be in the appropriate place to read the technologies. The "seats" of an experimental hall *theatron,* as it were, are in front of terminals attached to the detector via computers. The audience for scientific experiments does not consist only of the members of the experimental teams who enact them, but also of a community of scientists that has been suitably prepared through a training process which enables them to read the technologies. (Theoretical contemplation of experimental performances, which does not require firsthand observation but takes for granted accounts of their real presentation in the laboratory, are a different way of witnessing experimental performances, one that will be discussed in the next chapter.) This community witnesses the event through information disseminated by the primary experimental team in reports, colloquia, talks, preprints, journal articles, and the like. Such dissemination requires performances of an entirely different kind: oral presentations, and preprints and journal articles have their own style and rhetoric and can be treated as a subspecies of literary performance.

The Technology and Artistry of Experimentation

Within the laboratory one encounters readable technologies for the kinds of projects one wishes to execute in it. If one wants to obtain, for instance, the spectrum of a compound, the chemical composition of a material, or the electrophoretic separation of a substance, laboratory equipment provides ways that one uses for so doing. To be sure, passive forms of preparation may turn out to be not what one supposed, and subsequent events may lead one to reexamine and adjust them.

> *In January of 1990, researchers at the Mecklenburg County Environmental Protection Department in Mecklenburg County, North Carolina, who were testing substances for organic compounds in a mass spectrometer, began to get an unusual reading indicating the presence of benzene, a carcinogen. The materials tested, however, showed no trace of the substance. At first the lab suspected the instruments were malfunctioning, but tests showed they were functioning normally. Various materials and utensils that went into the testing process were examined for impurities, but they, too, proved clean. Eventually, the contaminant was traced to the source of purified liquid that the scientists had been using to dilute the specimens: Perrier water. Less newsworthy examples of anomalous behavior by the forms of passive synthesis crop up all the time in laboratories in the form of impure materials, defective cables, poorly machined parts, unanticipated factors, and the like. A few years ago, what had been thought to be surface features of DNA as it appeared in scanning tunneling microscopes turned out to be surface features of the particular kind of graphite substrate on which the DNA was frequently deposited.*[5]

But episodes such as these are exceptions; the process of experimentation itself requires that cases in which one has been misled completely by the equipment be exceptional, the stuff of lunch table anecdotes. For the ability to take the operation of a technology for granted is precisely what makes it "readable." This passive aspect can be referred to as the *technology of experimentation.* Advances in the technology of experimentation make possible greater performance capabilities. But one cannot separate the technology of the experiment from the end; the performance one wants to achieve. One has to decide the science one wants to look at, and then the performance conditions and signatures of that science, which may involve tradeoffs; everything cannot be optimized simultaneously. A negotiation may thus take place between what one hopes to see, what is likely to be seen, and the technology available for such seeing. This negotiation—a hermeneutical process—can be quite complicated in the planning of the large particle detectors of high-energy physics.

Active forms of preparation involve acts of artistry and judgment above and beyond the readable technologies in order to make phenomena appear. One can refer here to the *artistry* or *prudence of experimentation.*

A few philosophers of science have recognized the existence of artistry in science although generally without placing it within a larger vision of science. Gerald Holton writes of the "good run," which is reminiscent of actors' descriptions of a "good performance"; a skilled experimenter such as Millikan had the ability to judge whether or not a given run was "good" without knowing whether it confirmed the hypothesis on the table. Holton also writes of theorists' *Fingerspitzengefühl,* or instinctive feeling for the state of affairs, by which they judge the relevance of hypotheses.[6] When Michael Polanyi writes of "tacit knowledge," and states that scientists' personal participation in research is indispensable, he is edging toward the acknowledgment of an artistic dimension to science.[7] Evelyn Fox Keller's notion of "dynamic objectivity," involving a form of attention to the natural world that respects the kinship of organism, environment, and scientist, is more explicitly artistic. Injunctions from Barbara McClintock to "listen to the material," "let the experiment tell you what to do," and acquire a "feeling for the organism" emphasize the element of craft in science; Keller, McClintock's biographer, acknowledges her debt, in developing the notion of "dynamic objectivity," to Ernst Schachtel's description of how an artist focusses attention on nature.[8]

More common are those philosophers of science who ignore the aesthetic dimension of science entirely, or who pretend that it pertains only to theory. Amazingly, entire conferences and books have been devoted to the subject of "aesthetics" in science that focussed almost entirely on the role of aesthetics in theory and explanation with virtually no recognition of its (omnipresent) role in experimental activity.[9] One can only regard this as perverse, given that "aesthetics" derives from the Greek word *aisthesis,* meaning the power of apprehending things by using our senses, which was opposed to *nous,* or the power of apprehending intellectual and immaterial things (like theories and explanations). But in the mythic account,

aesthetics has come to be divorced from truth and opposed to the procedure of science. "The question of truth," wrote the turn-of-the-century
logician Gottlob Frege in one of his most famous works, "would cause us
to abandon aesthetic delight for an attitude of scientific investigation."[10]
But aesthetics, so far from being opposed to scientific procedure, is an
integral part of it, for experimentation involves bringing something materially into being through skillfully created actions, along with theoretical
investigation of ideal forms. While it is one thing to laud the economy and
artistry of individual experimenters such as Rutherford, Fermi, Monod
and others, it is another adequately to incorporate a role for that economy
and artistry in a philosophy of science.

The issue may be broached by a consideration of the difference between
four types of performance: (1) a *failed* performance; (2) the mechanical
repetition of a performance; (3) a *standardized* performance; and (4) an *artistic*
performance. A failed performance is one in which something whose appearance was anticipated did not appear, either through malfunction of
the equipment or some other reason. Nothing new or interesting appeared
in the mediocre *Hamlet;* nothing could be told from the poorly executed
experiment. A mechanical performance is exemplified by videos or player
pianos, whose encoding or program ensures that the same events or notes
will appear in the same way. These mere repetitions are only very loosely
speaking performances, for they are not creations but static echoes of
something; no uncertainty exists about the outcome. Demonstration experiments are like this. A standardized performance is one that does no
more than fulfill the standards of a performance tradition; it cleaves to
accepted practice, ventures no further, and is relatively oblivious to the
environment. In a standardized theatrical performance, for instance, one
acts as though the role were already constituted; standardization means
performing with a minimum of interpretation, synthesis, and artistry. The
performer can do what "one does" with it, or simply "phone in" the role.
In the case of scientific practices, what is involved in standardization is
similar but the outcome can be positive. Practices that only a *certain* scientist or scientists can execute successfully become transformed into practices
that a wider group of scientists without as much training (or even members
of the general public, with *no* training) can execute successfully. A standardized practice is one that has stopped being something that a *certain*
scientist can do in a community and is now something that *anyone* in that
community can do. Practices may be more or less standardized, depending
upon the amount of skill and/or training that one must have to use them
successfully. A classic description of the difficulties involved in the standardization of a scientific practice is described by Ludwik Fleck in *Genesis
and Development of a Scientific Fact.* In the case of the Wassermann reaction
as a test for syphilis, Fleck describes how the necessary serological skills
were developed by Wassermann and then painstakingly developed and

standardized sufficiently to be passed on, first to associates and then to other laboratories.[11] The test did not initially "travel well," but soon became mobile and thus effective. While standardization lacks originality and is often poor theatre (but not necessarily; it may be dramatic and exciting to enact a role *à la* Olivier or Welles), it is good technology; rather, it represents the transformation of science into technology.

Artistic performance, on the other hand, coaxes into being something which has not previously appeared. It is beyond the standardized program; it is action at the limit of the already controlled and understood; it is risk. The artistry of experimentation involves bringing a phenomenon into material presence in a way which requires more than passive forms of preparation, yet in a way so that one nevertheless has confidence that one recognizes the phenomenon for what it is. Artistic objects "impose" themselves—they announce their presence as being completely or incompletely realized—but this imposition is not independent of the judgment and actions of the artist.

> *I once concocted a scheme to repeat the famous and epochal experiment by which Ernest Rutherford discovered the atomic nucleus, which inaugurated the atomic age. In practice, it looked simple: a source, a target, scintillation screens, tiny flashes which you counted in the dark. I had even seen pictures and diagrams of the equipment. I envisioned performing it in front of students—even making a video or documentary perhaps. To assist me I approached someone whom I knew had worked with Rutherford on alpha scattering experiments. My suggestion made him roar with laughter. After he finally quieted down, he explained to me that obtaining the permits to work with radioactive material of the requisite strength was virtually impossible today. Then he said: "The main problem, though, is that experiment is a craft, like making an old violin. A violin isn't a very complicated-looking gadget. Suppose you went to a violin maker and said, 'Could you kindly help me make a Stradivarius? I'm interested in violin making, and I'd like to see how it was done.' He'd smile at you just like I did. Because craft is a knowledge you have in your fingertips, little tricks you learn from doing things, and they don't work and you do them again. You have little setbacks, and you think, how can I overcome them? And then you find a way. Every time your equipment changes you forget all the old techniques and have to learn new ones. And you have to know then, because when you're pushing your equipment to the limit it's bloody easy to get spurious results. You're scratching at the ground all the time, and you don't know what you've missed. Every experimenter has made terrible errors at one time or another, and knows of instances where friends have fallen on their faces because they got spurious results and published too early. And yet, you've got to push what you know to the limit. If you don't, someone else is going to do it first. And that's dreadful, being beaten. Everyone's got a*

closetful of discoveries they missed because they were too cautious or some other fellow was cleverer. There was a whole Austrian school working on the same things as Rutherford at about the same time, and nobody's heard of them today. Why not? Rutherford was just a little more daring and crafty."[12]

Like artists, experimenters are restricted by the limits of their equipment and materials; they push these limits and must wait and see what works. Were the experimenter in complete control of the action, there would be no experiment, for the result would be known ahead of time. The experimenter notices the action of the equipment, the ways in which this action cannot be subdued, the ways it resists attempts to alter it. When this happens, something is exhibiting different profiles in a structured way: what provides the law governing the sameness of what shows up in different appearances, we know thanks to Husserl, is the invariant. By running the "performance" again and again, you do not get different profiles; when you run it differently, you get different profiles, but in a lawlike way. Such resistances to transformations, showing the presence of an invariant, are hardly manifestations of the imperfection of nature but of its *presence*.

An artistic performance begins with a performative play-space that is not infinite. A performance is fresh and unique when it is synthetically attuned to the specific conditions of the environment in which it takes place. (A standardized performance transpires in relative detachment from the surroundings.) I say synthetically because the performer acts as though stage, audience, lighting, the other performers, and all the details no matter how tiny were not things in space, not simply there and copresent, but agents operating on the performance: coperformers. A performer allows such things to function as organic parts of the performance as event.

For experiments, like performances of any kind, are *holistic* in that they involve the simultaneous working together of a number of different kinds of elements. In a good performance one cannot draw a line between essential and inessential aspects, and even the tiniest detail can require supervision. Experienced performers know one can never predict what will make or unmake a performance. Experimenters are also aware of this lesson, through numerous wrong experiments in which failures to supervise apparently inessential details spelled the difference between success and failure. The pseudodiscoveries of polywater and, more recently, cold fusion are cases in point. Small things like dust, iron filings, overhead lighting, temperature sensitive detectors, impurities, and the like have doomed hugely complicated and otherwise well-supervised experiments. One could also say that performances are anaphylactic, or hypersensitive to ambient conditions. Each artistic performance, rather than repeating or echoing, is a creation that pushes forward to produce what is *repeated*.[13]

Attaining such a creation requires a capacity for something like play.

Play here can be defined as "the performance of the movement as such," an action that transpires for its own sake.[14] Play, like performance, absorbs and transforms the players. This does not mean that the movements are predetermined. Just the opposite; play involves the presence of possibilities and freedom of action, and the presence of risk is constant. To perform (rather than to imitate a performance of) *Julius Caesar* is not to repeat the gestures of previous performances, but to arrive at new gestures that present the work appropriately in the world in which it is performed. The work reveals its different profiles not through repetition but by manifesting inexhaustible differences. Putting this another way, performances are *probative* (exploratory) in that what takes place is not fully predictable in advance. One does not know precisely what will occur if one is truly performing rather than rehearsing, calibrating, or demonstrating; there is an aura of expectancy and suspense when a good performer takes the stage, even at the end of a run. We may also express this by saying that the act is executed in response to an inquiry, taken in its broadest sense as that "vague fever" that Merleau-Ponty asserts is prior to the act of artistic expression, a fever that cannot be assuaged with the aid of books, theories, and the like, but only from what transpires in the interplay of the performance elements: "[O]nly the work itself, completed and understood, is proof that there was *something* rather than *nothing* to be said."[15] Only the experiment itself, completed and understood, is proof that there was something rather than nothing to be discovered.

Moreover, the play has no end but itself, and consists of a backwards and forwards motion between the players and something that is "played out" through them. Each game is, as it were, a singular profile of *the* Game, the exemplary game. That something, the Game, that gets played out is what is important here. In the experimental situation, the *play* takes place through the experimenters on the one hand, and the actions of the equipment on the other. The backwards and forwards motion takes place as the experimenters adjust their equipment, responding to what is being "played out," to make what appears do so more clearly. What gets played out in the experimental situation lies beyond the initial program, and when successful can involve the appearance of a new phenomenon or of a new profile of a familiar phenomenon. How something appears depends upon the horizon of skills, techniques, and concepts possessed by the community. What one might call the *sculpting* of the experimental phenomenon—the adjustment of the equipment to make the phenomenon appear more boldly, to throw it in more relief—will be affected by the state of the background context. The anticipatory goal is the appearance of the phenomenon. But especially in the case of a genuinely new phenomenon, one's anticipations of how it appears emerge together with the sculpting.

Performances are also *authoritative* in that they demand acknowledgment by those engaged in inquiry into the sort of activity being performed; there is always already a relatedness between the act and those witnessing the act. What appears in performance does so not magically, not as an exception to the order of the world like a magician pulling a rabbit out of a hat, but from an observed coworking of elements. One can see how it is done; nothing is concealed; it is the opposite of magic. However virtuosic or extraordinary, a performance is still a worldly event which fascinates because its presence must be acknowledged in worldly terms. This is the case even when what looks or is asserted to be an appearing phenomenon turns out to be an epiphenomenon, for what then imposes itself is the demand to explain the epiphenomenality appearing in performance. It was not enough for the scientific community to *disbelieve* the experiments that purported to have demonstrated cold fusion; one insisted on knowing *how* they had gone wrong. An experiment may seem cryptic and recondite from the outside, and be accepted by outsiders on faith—taking the word of what "they" say—and skepticism can also be voiced. But the true audience is a trained and specialized group whose members know how to *read* the actions of the performers. (The same is true for theatre, which is usually not intended for a real cross section of society.) The audience knows how to make sense of what is happening in principle even if in fact it takes much of what transpires on faith.

Text and Act Hermeneutics

The back and forth motion between experimenter or performer and equipment in the emergence of a phenomenon suggests that this is a hermeneutical process for which our discussion of Heidegger and interpretation is appropriate. In chapter 2, I introduced Heidegger's conception of the hermeneutical circle in the interplay between a background knowledge and experience and the world. This interplay involved three moments which are implicated in each other, a *Vorhabe* (forehaving), *Vorsicht* (foresight), and *Vorgriff* (foreconception). To review briefly: a *Vorhabe*, or forehaving, is the set of culturally and historically acquired involvements that we "have" in advance without which there would be no inquiry or problematic situation; the *Vorsicht* is the vision or anticipation of what would resolve the problematic situation; the *Vorgriff* involves an understanding of how to initiate the reconstruction. If my inquiry, say, involves attempting to produce a certain sound with my violin, I simultaneously have an ability, however rudimentary, to produce notes, an idea of what kind of note I am trying to produce, and a recognition of how to go about trying to get the kind of note I want based on my abilities. In learning to play the violin, I put into play all three moments all the time, developing, deepening, and enriching my interaction with the instrument.

Commentators, however, do not usually elaborate the hermeneutic circle with examples such as the above, but instead with examples of textual interpretation, or the understanding engagement with texts. What is involved here, however, is not *text* but *act hermeneutics*, or an understanding engagement with the world in the performance of actions. Act hermeneutics also involves the hermeneutical circle and can be examined accordingly.[16]

It is no coincidence that the hermeneutical circle is generally associated with textual interpretation, for artistic skill, being at a remove from cognitive behavior, is difficult for philosophers to address. When philosophers have spoken of artistry at all, they often attribute it to blind inspiration, the inexplicable presence of the Divine, as did Plato in the *Ion*. But the example of theatre suggests a way of applying the hermeneutical circle to artistry. The *Vorhabe* can be thought of as involving the body of knowledge that a performer brings to the role or performance opportunity. But as C. P. E. Bach's friend, the noted teacher Johann Quantz, liked to say, "However well ordered the fingers may be, they alone cannot produce musical speech"; another moment in the learning or skilled execution of a performance is a vision of what one wants to produce.[17] The third moment involves knowing where to turn to work toward achieving that vision.

All three are at work simultaneously in the acquisition and execution of performances in which something is materially brought into being. Consider theatrical rehearsals. "In rehearsal," writes director Joseph Chaikin, the actor "exposes himself to the elements which together form the event." Drawing on the actor's particular resources, the actor adapts them to the particular circumstances of the performance. Chaikin continues, "The actor must come to a connection with the material as a person is connected with his environment."[18] Rehearsal aims to intensify and coordinate the performer's understanding and skills, bringing them to bear on the specific requirements of the piece. Piano teacher Mildred Chase even makes her students take account of the effect of humidity on the operation of the piano, and insists that they be aware of how the ridges on their fingertips affect their experience of the instrument.[19] The result both transforms the abilities of the performer and brings the role to presence.

Experimental performance shares the same essential structure. In any inquiry, experimenters possess a background set of skills, practices, and other commitments without which there would be no inquiry. But any problematic situation also involves a question that involves this background knowledge, as well as an intimation of how to use it in reconstructing the question. The movement of these three moments characterizes the skill in judging to which Aristotle refers as *phronesis*, prudence or practical wisdom, or the ability to act excellently with regard to human goods.[20] Aristotle divides the kinds of knowledge involved in material activity (whose objects can be otherwise) in two: craft or *techne*, and practical wisdom or *phronesis*. By *techne*, Aristotle means the kind of activity involved in production, such as throwing a vase or building a ship. It is like *episteme* (pure or disinterested contemplation of general rules or principles whose intellectual process is demonstration and whose objects can be otherwise), in that the end is fixed and determined in advance— the idea of the vase or of the ship. Unlike *episteme,* the actual object, the outcome of the activity, can vary. The distinctive kind of knowledge involved in experimentation is not *techne,* inasmuch as its end, the performance itself, is not determinable in advance. Operating an accelerator for medical or industrial purposes is not an act of experimentation but of engineering or technology. Lab exercises such as those performed in physics courses are also not experiments; they are demonstrations for the purpose of illustration or training.

The distinctive kind of knowledge involved in experimentation most resembles *phronesis* or practical wisdom; being wise in the world through activity. *Phronesis* is the activity characteristic of a statesman or judge, an activity whose end—the good, just, or right action—is not determined in advance. A judge applies laws, but laws are general and each actual human situation is different from every other. Laws must be constantly adapted and adjusted in practice, not because they are imperfect, but because of the innate complexity of cases to which they are applied. Because the end of such activity, the good act, is undetermined beforehand, the intellectual process involved is not deduction or production but deliberation. As Aristotle is fond of saying, no one deliberates about things that could be other than they are. In the case of experimentation, read "good or just result" in place of "good or just action." The deliberation involved in experimentation occurs not on the level of reading dials, but of planning, construction, and interpretation.[21]

Experimenters employ skills other than *phronesis.* A contemporary scientific experiment involves many different kinds of tasks that require many kinds of practices, techniques, tools, and equipment. Carrying out these tasks often involves constructing well-understood equipment and applying or repeating standardized techniques and procedures. But such skills are also those of the technician or engineer. What is distinctive about the experimenter is performance, where the operation of the equipment is a matter of artistry or prudence.

Specifying the kind of skill characteristic of the experimenter is important for three reasons. First, it keeps us from the temptation of thinking we can write a definitive set of rational experimental procedures. This would be like trying to write a definitive set of rules for judges to follow to guarantee just decisions, or for actors to follow in skillfully performing roles; if it were possible, it would have been done already. Shall we claim that it ought to be possible to write a second set of rules for applying the rules of theory, such as a set of "rational experimental procedures"? But in that case, as Kant observed at the beginning of the Analytic of Principles in the *Critique of Pure Reason,* a third set would have to be written to describe how to use the second set, a fourth for the use of the third, and so on. Ultimately, one is thrown back on the need for an ability to apply rules that is not itself rule-governed.

Second, to specify the kind of knowledge employed by experimenters is to understand the nature of so-called missed discoveries and pseudoeffects. History is full of examples of good and careful scientists (Fermi, the Joliot-Curies) making what, in hindsight, are misjudgments. The reason is that the application of *phronesis* inevitably involves an element of risk. Deliberation can be based on missing or incorrect information, or it can simply misjudge the contributing effects.

> *In 1957, theorists were piecing together a picture of the weak interaction, the force responsible, for instance, for beta decay. But the attempt was frustrated by experimental evidence showing that, while the character of the transformations of a particle's wave function during an interaction is in most cases a mixture of V (for vector) and A (axial vector), specifically V − A, the results of four experiments indicated that it was something else, including one involving He[6] indicating that it was V and T (tensor), performed under the aegis of Chien-Shiung Wu; her experiment less than a year before had been the first to demonstrate the existence of parity violation. But at a conference in Padua, Italy, in September of 1957, physicist Robert Marshak, convinced of the universality of the character of the weak interaction, boldly declared that these contrary experiments had to be wrong. "I said that it had to be V − A for a universal interaction, and that to be so will require that four experiments be murdered—I used the word murdered—including the He[6] experiment. One physicist then said to Lederman during a coffee break, 'Marshak's crazy. How can He[6] be wrong?'" But within months, a carefully performed experiment revealed that V − A was indeed correct.*[22]

Without an idea of a kind of knowledge such as *phronesis* at our disposal, which involves deliberation and therefore risk, we are forced to try to ascribe all such examples of wrong experiments to error or oversight on the part of the experimenter rather than to the innate ambiguity of reaching a good result. If a way to eliminate risk from activities involving *phronesis* were possible, experimenters would never err, virtue would have triumphed in the courtroom long ago, and the performances of actors would all be great.

Third, specifying the kind of knowledge characteristic of the experimenter is important in underscoring the need to examine the practice of scientists. Experimental work is not just the implementation of theory. It has its own distinctive kind of knowledge, its own distinctive kind of achievements, and its own distinctive kind of risks.

The role of artistry and phronesis in the preparation of scientific entities is absent from the mythic account of science and experimentation. The weight of the traditional philosophy of science is set against recognizing their presence and necessity in science. The distinction between *context of discovery* and *context of justification*, for instance, conceals the role of artistry and prudence, holding them to be finally irrelevant. But the schema of act hermeneutics allows them to be recognized and assigned their proper place, and provides for an understanding of why they have been misunderstood. For through standardization phenomena revealed first by artistry can come into presence in a routine way (in science museums, for instance, where the push of a button is sufficient to make a profile of a phenomenon appear), apparently dissociated entirely from human actions and certainly from artistic ones. A prize example is piezoelectricity, the phenomenon that certain crystals, when squeezed in certain ways, produce momentary jolts of tens of thousands of volts of electricity. The phenomenon first put in a laboratory appearance around the turn of the century through the artistic work of the Curie brothers in complicated laboratory equipment. By the time of World War II it had been sufficiently standardized to use in the nose of aerial bombs in detonation devices. Standardized even further, that once exotic laboratory phenomenon is today a commonplace feature of the ignition systems of certain kinds of cigarette lighters, one of our culture's emblems of disposability.[23]

But the real appearance of scientific phenomena cannot be separated from the forms of preparation, active or passive, by which they achieve presence. Only through performances, standardized or artistic, are they parts of our world, are they *there* in the laboratory rather than not. (They can, of course, have manifestations outside the laboratory.) The appearance of the phenomenon cannot be separated from the environment in which it appears, except through the abstraction or representation of theory. The phenomenon must belong to the world in which it appears, and artistry can be involved in making it appear in ways it has hitherto not.

This oversight of artistry and prudence has been facilitated partly by scientists themselves, who frequently hold themselves and their colleagues to an exacting, uncompromising, unsentimental, and ultimately even unrealistic standard regarding responsibility for the quality of scientific work.

Nobel laureate Leon Lederman, for instance, the former director of Fermilab, the national laboratory in Batavia, Illinois, likes to chide himself over his "missed discoveries," and once wrote a paper about what he later called "the big ones that got away."[24] *Lederman counted among these the time that a team of which he was a member nearly landed an important particle that six years later was simultaneously discovered by two independent teams of researchers unconnected with Lederman. "Our thinking," Lederman wrote, "[and] our grasp of the crucial elements of the physics, were fuzzy."*

But the work of Lederman's team was treated by colleagues as first-rate, and both teams that finally discovered the particle, called the J/psi, used the results of his team as a guide. On one occasion, Lederman was asked whether he really believed that he lacked a sound grasp of physics in that experiment. "It wasn't sound enough," he replied. The objection was put that his experiment was, and still is, regarded by colleagues as a wonderful experiment. "Not wonderful enough. If it had been a little more wonderful, we would have found the J/psi. I should have been smart enough to use fine-grained detectors." Reminded that he had used certain thick materials that precluded the use of that kind of detector, Lederman shook his head obstinately: "I should have been smart enough to take out the thick materials and put in thinner ones." But, continued the protest, that would have meant changing the entire scientific goal and physical structure of the experiment on highly speculative grounds. Lederman was unmoved. "If I had been smarter," he brooded, "I would have started that experiment over from scratch. But I wasn't. I was dumb."[25]

Why do Lederman and other scientists habitually and insistently adopt a self-deprecating attitude about their efforts, and refuse to acknowledge the artistry and prudence in it? Their attitude, a convention that defines what it is to have "the right stuff" in science, attributes all failure to poor planning and judgment, and denies the inherent risk and uncertainty in experimental efforts. This attitude inspires them to greater effort in their risky and demanding work, forcing them to offer no excuses for themselves or others. A philosophical account, however, must not be taken in by the romantic account manifested by such an attitude—however appealing and worthy of emulation it may sound. A philosophical account must be able to see behind the account, explain its origins, and offer a description of the genuine artistic process cloaked by it.

The primacy of performance here means that the technology and artistry of experimentation are subordinated to the appearing of the phenomenon in performance—to making the phenomenon appear more boldly, to throwing it into greater relief. The performance is *in the service of* the phenomenon, the way that interpretation for Heidegger is, one might say, in the service of the meaning that appears. The appearance of the phenomenon, as an event, is then equally present both to players and audience. It is incorrect, then, to speak of a theatrical performance (or, for similar reasons, a sports match) as taking place "for" the audience. The performance is an event itself, something that "plays itself out," and those who sit in seats around the stage are incorporated into that event as surely as those who actually participate in it. The performance "takes off" when the members of an audience fade into the background as spectators and acquire presence as participants. Gadamer remarks that "the difference between the player and the spectator is removed here. The requirement that the play itself be intended in its meaningfulness is the same for both."[26] Theatrical and sports players who intend not the meaningfulness of the game, but the adulation of the fans, tend to be poor performers and their talents suffer. For the game is not aimed at them but at the performance. The meaningfulness of the performance is preserved, Gadamer says, even when a performance is held without any spectators—for the players only— as it is, for instance, in chamber music: "If someone performs music in this way, he is also in fact trying to make the music 'sound well,' but that means that it would be properly there for any listener. Artistic presentation, by its nature, exists for someone, even if there is no one there who listens or watches only."[27] The same structure appears in scientific experimentation. The performance—what is enacted by the equipment—involves the appearance of a phenomenon, and what enters into it is in the service of that appearance. When the phenomenon—electrons, let us say— appears in the experimental performance, that phenomenon is equally present to those (producer-director) scientists who actually have a hand in executing the act as to those who merely look on. Experiments thus take place no more "for" the experimental groups who enact them as opposed to an audience formed by the community at large than theatrical performances take place "for" the audience in the hall as opposed to the players. True performance of whatever sort absorbs players and audience in one comprehensive event, an event dominated by the appearance of a phenomenon.

One does not turn on the equipment and—voila!—a revolutionary new discovery. Generally, the presence of the novel or unexpected in an experiment perplexes an experimenter, who may respond by adjusting the equipment and introducing variations. The experimenter may still be confused, but will have a better sense of what is confusing and what additional steps might be taken to alleviate the confusion. The anticipatory goal is the appearance of the phenomenon. But especially in the case of a genuinely novel phenomenon, one's anticipations of how it appears emerge together with the shaping of the experiment, often changing in the process. One therefore "gives oneself over to" the performance the way one gives oneself over to a game; one allows oneself to be surprised and transformed. The documentation of the theatrical arts is full of examples of how phrases or characters that turned out to be archic in the final versions of works—the major third interval at the beginning of Beethoven's *Symphony No. 5* in C minor, the role of the Marschallin in Strauss's *Rosenkavalier*, for instance— appeared in early drafts less clearly defined and in subordinate positions. The history of science, too, is replete with stories about how apparently minor, irrelevant, and irksome details—glowing screens, scatterings of particles, leaky electroscopes—grew from nuisances to key features of experimental performance, sometimes over decades. The artistry itself is subordinate to the appearing of the phenomenon; the artistry is artistry *in the service of* the appearing of the phenomenon. This is despite the fact that such artistry could become the focus of an inquiry or a paean in its own right. To recall Dewey's point about the primacy of the successful production that ensues from inquiry over both theory and practical knowledge, each of which derives its significance from such production, one might say that skill divorced from it is tedious and repetitive, while theory separated from it is arbitrary and whimsical. For artistry reveals itself as artistry truly in performance; that is, when connected with the appearing of phenomena. Elsewhere, it is artificiality and "mere technique."[28]

Finally, the artistry of experimentation, like that of theatre, is often accompanied by a feeling of joy and celebration. This feeling seems naturally to accompany the bringing of a novel phenomenon materially into being in such a way that one nevertheless has confidence that one recognizes the phenomenon for what it is. When scientists are not on guard about presenting the image of being detached observers, they, too, can voice expressions of the enchantment of creation similar to those often found in memoirs by dramatic artists referring to the experience of giving birth to a new experience that is nevertheless recognizable. Moreover, a related sentimental attachment can develop over instruments or laboratories, similar to that of actors to locations, plays, or collaborators. One looks in vain for a place for or even a recognition of these kinds of emotions in traditional accounts of the philosophy of science. In fact, philosophers of science such as Frege have gone so far as to oppose aesthetic feelings to those that accompany genuine scientific inquiry. But those who consider such emotions irrevelant to the process of experimentation and scientific inquiry have not fully grasped the kind of activity that science involves.

> When Brookhaven National Laboratory, in Long Island, New York, dedicated its first major particle accelerator, the Cosmotron, the first machine to accelerate particles past 1 GeV, the ceremony was quite a symposium. "At least one guest passed out on the table, and a Berkeley scientist set his tablecloth on fire. The final speaker, Dr. Detlev W. Bronk, the president of Johns Hopkins University, mixed up the text of his speech with one he was scheduled to give in Canada, puzzling those still compos mentis with references to 'your King.' Because of the magnitude of the accomplishment, no one found the revelry excessive. 'A billion volts?' [accelerator physicist John] Blewett later remarked,'—that was one helluv'an achievement!'"[29] Sentimental moments have also accompanied the shutting off of machines— including the Cosmotron, after fourteen years of service.[30] Are moments such as these incidental to what science is all about? True, they appear only infrequently in histories of science, and are almost never taken up by philosophers as a subject of inquiry. But one finds them if one knows to look.
>
> The joy of creation is all over the pages of The Double Helix. And in an interview regarding an extremely difficult experiment conducted in the mid-

1970s concerning parity violation in atoms, the severe and reserved physicist Richard Taylor (later a Nobel laureate) quietly but firmly expressed precisely that joy in terms scientists can easily recognize: "I remember that in just three days you went from wondering whether the experiment was going to work to being pretty sure that you knew that there was parity violation. That, I mean, that's why you do this business. That feeling of knowing something before anybody else. It's, ah, it's why you're here."[31]

V

PERFORMANCE
REPRESENTATION

Theory, its nature and structures, are the subject of an extensive philosophical literature; is there any textbook in the philosophy of science without at least one chapter on theory? This is a topic that has been "covered"; it is not one of the "holes" in our periodic table needing to be filled.[1] Therefore, I shall discuss it only in terms of the relation between it and the schema I have developed. But I shall also discuss two topics, the prominent role of mathematics in theories of the natural sciences and some peculiarities of quantum theory, that pose recurrent problems to the philosophy of science, to see whether this schema offers any contributions.

In the last chapter, I showed how experimentation, when successful, involves the *presentation* of phenomena that show themselves as "the same" (and not, say, as glitches, errors, or "artifacts of the machine") to firsthand observation by readable technologies through different laboratory procedures. I discussed various (philosophical) means of understanding what is involved in such presentation. In this chapter, I address how such phenomena are *represented* through theories. Theory formation does not require firsthand observation of phenomena but takes for granted accounts of their real presentation in order to represent what it is that remains "the same" in different laboratory appearances.

I shall (1) sketch out the role of theory in the schema already developed, of successful experimental acts as a species of performance that produces profiles of phenomena as part of a process of inquiry, and utilize the theatrical analogy to point up certain special features of theory overlooked by traditional accounts; (2) discuss what light this sheds on one of these special features in particular, the "unreasonable effectiveness" of mathematics in the natural sciences, which is due to the privileged role of group theory in representing transformation groups; and (3) discuss the difference between classical and nonclassical phenomena, and problems involving the understanding of the nature and ambition of science (and in particular quantum mechanics) created by this distinction.

Theory as Scripting

Every performance is related to some kind of representation, or program such as a text, score, or script.[2] This representation can then be used as a program for the performance of further experiments to explore that phenomenon. A program, of course, can itself become an object of knowledge; Maxwell's equations or Newtonian mechanics or the theory of relativity can be treated as a field of study in its own right and found to contain discoveries, surprises, and even inconsistencies. But the body of knowledge that emerges from such study no longer concerns things in the world strictly speaking—phenomena—but a field of their possible profiles. Thus, theoretical knowledge is but one aspect of scientific knowledge, for science aims not only to probe possibilities, but also to show actualities.

The role of theories, models, and other forms of representation, we saw in chapter 3, can be considered in two ways. In one, they are treated as coherent and independent systems that picture essences beneath the surface of the world—"scientific images," in Sellars' phrase, as opposed to "manifest images." This is the way traditional philosophy of science has for the most part approached not only theory but science itself, assuming the presentation and recognition of performances to be unproblematic. In the classical understanding of theory, for instance, phenomena are pictured by an abstract Hilbert space (a kind of vector space) of ket-vectors, each with a unique, precise value, which is the limit to which perfectible measurement processes tend.[3]

In the other way of considering the role of theories, models, and other forms of representation, they are treated as ways of organizing perceptual information about the world. In the schema I am developing, the perceptual information consists of profiles of phenomena. The theory and the appearance of profiles of phenomena it represents are linked by instrumental praxes.

A theory, we might say, *scripts* a phenomenon. But a script is related to a phenomenon in two ways. It structures the performance process (it "programs" the performance) and it structures the product of the performance. A theatrical script, for instance, both structures the actions of the performers on the one hand, and describes what is heard (the play performed) on the other. Read noetically, a script is something to be performed; read noematically, it describes the object appearing in performance. In the context of experimentation, however, what is presented may show itself in such a way as to suggest another manner of scoring, so to speak. The act of experimental performance, while controlled by a representation, is done with the possibility in mind that what appears may do so in a way calling for a new representation. When a theory *works*, the model fits what becomes present in experiment, and what becomes present fits the model. Expectations match what is disclosed, and what is disclosed fulfills expectations. These expectations are structured by the instrumentation, and the model helps us understand what that instrumentation is *reaching*, or presenting to us. For the instrumentation does not present us with the model, but rather with some thing which through inquiry is seen to have (or to lack) the structures of the model. The model of the double helix allows us to move from profile to profile in the laboratory, instructing us in our manipulations and helping us to understand what is given through them.

In music, we would consider a successful score to be one that, on the basis of standard practices, would allow us to reenact the event. What would be considered to be adequate scores (notational practices) in one period of history would not necessarily be considered to be adequate in another period, when different kinds of music prevailed, played in different ways on different kinds of instruments. A score does not picture or represent some ideal performance. Even composers such as Mozart, who reputedly had the ability not only to recreate a symphony in his imagination upon reading a score but also to *write* one in his head as well, relied on standardized practices (instrumental and notational) to do so. The *Noten im Kopf* to which he was listening in the privacy of his internal theatre were played on a clavichord, an instrument of his day, rather than on a pianoforte, an instrument perfected in the next century, or on a synthesizer, an instrument of our time. And Wilshire points out that in imagining their dramas performable, playwrights do so "within an already-standing tradition and institution of performance."[4] If Mozart's symphonies are performed on synthesizers, or if plays are performed in ways that break with traditions and institutions, then the scores or scripts have not been violated; they remain the scores or scripts for those events. Scores or scripts are open-ended regarding possible performances. And for many playwrights, for instance, there is a limit to what the imagination can disclose about performances; there is no substitute for witnessing what happens in performance, and revisions almost invariably follow readings, rehearsals, and previews in order to shape better what appears in performance.

Likewise, theory cannot be viewed as picturing entities that exist apart from the life-world; from particular standardized means of perceiving the phenomena. A theory represents a phenomenon by scripting or scoring the performances in which it appears. It represents an invariant which through standardized techniques and practices can be correlated with elements and operations of performances. A failed, inadequate, or incomplete representation is one that cannot be correlated with the elements and operations of a performance; it may indicate profiles that do not appear in performance, or may fail to be correlated with the entire set of what are evidently profiles of the same invariant. And theory making alone (along with all methods of investigation that rely heavily on imagination, such as Husserl's method of profile variation) cannot provide genuinely new information about phenomena, but only about possibilities for how already anticipated phenomena might appear; moreover, one still has to check whether the representing model provides a valid description of the represented phenomenon.

Theories in science seem different from representations of music, dance, and drama because the former are constrained by (what is experienced in) the world. Although representations in these other art forms are also based on attention to experience, they nonetheless are viewed as the unique province of individual artistic imaginations addressing the social world, whereas scientific theories are thought to arise from a communal and objective encounter with the natural world. Had there been no Shakespeare or Beethoven, one may say, we would have no *Hamlet* or *Choral Symphony;* had there been no Maxwell we still would have something similar to the set of equations that bears his name and that have been used ever since their discovery.

Maxwell's equations indeed have been used in widely different social, historical, and technological contexts for over a century now, and at first glance would seem to have the character of an immutable natural law. But our understanding of how these equations are *applied* to the situations at hand—hence, their function in a *theory* of electromagnetism—has changed with the context, which now includes, for instance, quantum mechanical situations. Furthermore, Maxwell's equations themselves have had a considerable history, and have changed markedly in how they are written down and in other respects.[5] As I mentioned in chapter 3, they also have problematic features that make it likely that in the future they will be further revised.

Thus what we need for an adequate account of theory is not only the Husserlian notion of invariance, or regularity of profiles under transformations, but also the Heideggerian notion of interpretation, for what fulfills or "counts as" a profile is a hermeneutical process and always developing. The assuring, enriching, and deepening of involvements with nature changes both the invariants able to be picked out as well as the profiles that count as fulfilling them. Discovery, say, of a spectral line in the predicted location might count as confirming the theory of quantum electrodynamics at one time—but at a later time, with more advanced instruments and more refined theoretical understanding, discovery of a slight discrepancy between the original prediction and the actual location might *also* count as confirmation.[6]

The meaning of the primacy of performance for representation is that theory-making involves representation *of* performances; all theory is theory *of* performances of possible events. The development, elaboration, reworking and discarding of theories are all governed by how well they represent the invariances exhibited by sets of profiles in performance. The same is true in the dramatic arts, of course. Scripts and scores are not inviolate and sacred, but in actual use are changed to fit what works in performance.[7] Rightly used, scripts and scores serve the performance. Even the electronic information digitally encoded on a compact disc recording, a mass-produced replica of a performance that may seem to have a degree of objectivity and abstraction, has been adjusted in the process of mastering and production to make the music sound good when performed on the equipment on which the CDs are likely to be played. Hence what I referred to, with some exaggeration, in chapter 3 as the "fragility" of theory; it derives its value from its capability to program a performance, and is thus subject to change when the phenomenon shows itself appearing in a new way in performance, as well as with new means of performance. This is the sentiment behind Thomas Huxley's statement that "in science there's nothing more tragic than the slaying of a beautiful theory by an ugly fact."[8]

> In the mid-1970s, the theoretical movement toward unification of the weak and electromagnetic forces was stymied by experimental evidence of the lack of a parity violating effect when polarized electrons are scattered off atoms. Some theorists responded by developing a crop of "left-right" theories that preserved parity symmetry to the weak interaction. "[T]hose left-right theories just grew—I mean, there were acres of them, like spring flowers," said physicist Richard Taylor of the Stanford Linear Accelerator Facility (SLAC). Later, when a group co-led by Taylor concluded a carefully performed experiment revealing the existence of parity violation in such circumstances, scientists spoke of the "SLAC massacre" of left-right theories.[9]

Failure to fulfill a profile can also mean merely that unknown factors are causing a phenomenon to behave in a different way than expected, as may be the case with the lack of observation of the expected number of neutrinos emitted by the Sun, or the billions of tons of "missing carbon" that were put into the air by fossil fuel burning but which cannot be found in the atmosphere or accounted for by absorption in the oceans or by other means. But only through the performance can one find that something rather than nothing is *there;* only in performance does something enter the world. A script or score does not by itself testify to the presence of a theatrical performance, and even a performance based on such a script or score does not necessarily constitute a genuine theatrical event or phenomenon—it does not necessarily have any *presence.* The performance cannot be separated from the representation through theory, as if it were something inessential to the theory. And yet there is a deep ontological prejudice—Wilshire calls it "intellectuals' most basic prejudice"—that meaning lies in "writable meaning" or representation.[10] Rightly understood (to paraphrase Wilshire), theories describe the noetic-noematic structures of the encounters with human beings with nature via the mediation of instruments. These structures, and the theories that describe them, will evolve over time, however slowly.

The Role of Mathematics

Eugene Wigner once wrote an essay entitled, "The Unreasonable Effectiveness of Mathematics in the Natural Sciences," which concludes with an admission of defeat in his attempt to comprehend this effectiveness: "The miracle of the appropriateness of the language of mathematics for the formulation of the laws of physics is a wonderful gift which we neither understand nor deserve."[11]

In the natural sciences, scripting or representation often (but not necessarily) has a mathematical form. Theories can script performances via a variety of means, from general principles and outlines of programs for understanding to complex sets of equations, depending on the nature of the field and its state of development. The point is to describe the law of invariance of the profiles in a given outer horizon. As many experiments in geology, biology, and paleontology reveal, the presence of mathematics does not make one kind of theory intrinsically more "scientific" than another. Nevertheless—and this is Wigner's point—mathematics does play an exemplary role in some branches of science, and some kinds of theory construction. This role has been poorly understood, even by those skilled in developing and utilizing theories, such as Wigner himself.

It would indeed be a miracle, were there a Divine blueprint of the Universe out there that mathematics just happened to describe in the optimum way. But Wigner's own work, which did much to expand the recognition of the role of symmetries and invariances in the physical world, contains clues to unraveling the secret behind this apparent miracle. This clue is that the perceptual objects of the physical world frequently show themselves to us as having invariants that are invariants of transformation groups, and mathematics has a privileged role in representing such groups.

Many of the physical transformations that I can perform on a physical object, for instance, have a group structure, and these necessarily entail the presence of invariants. If I take an apple and turn it around, I obtain different profiles, but the profile revealed by each turn is related to that revealed in any other turning, so that the profiles have a certain order. I can turn the apple in several different ways in going from one given profile to finish up with another; these ways are all related and have a certain structure. Thus the spatial description of an apple (and indeed of any rigid body) includes the fact that it is the invariant of transformations of groups of rotations and translations. One need not use groups in the definitions of ordinary perceptual objects, because our ability to manipulate them embodies these structures. But groups do become important in definitions of certain kinds of scientific objects, which are presented in programmed performances of various kinds, for groups are an important means for representing the most general structures of physical objects that appear in these sometimes widely diverse performances.

> The Theory of Groups is a branch of mathematics in which one does something to something and then compares the result with the result obtained from doing the same thing to something else, or something else to the same thing. . . . [I]t is the most powerful instrument yet invented for illuminating structure. . . . Group theory has also helped physicists penetrate to the basic structure of the phenomenal world, to catch glimpses of innermost pattern and relationship. This is as deep, it should be observed, as science is likely to get.[12]

The development of group theory can be traced back to Renaissance studies of perspective, involving the relation between, say, a perceived three-dimensional object and the image of that object projected or mapped onto a two-dimensional canvas. It was noticed that despite the distortions necessarily introduced in this projection, certain geometrical properties remain unchanged; such properties are invariances or symmetries. The changes used in the projection or mapping between object and invariance were called transformations, and it was realized that transformations could consist of a variety of operations, including substitution, application of a rule or function, and change of coordinates. A systematic study of projective geometry did not begin until the end of the eighteenth century at the École Polytechnique in Paris. French mathematician Évariste Galois first used the term *group* in a technical sense in 1830, and further work on groups was carried on early in this century by his compatriot Élie-Joseph Cartan. Group theory was an important feature of the Göttingen school; Felix Klein's Erlanger Program, for instance, sought to classify all the various branches of geometry by the different kinds of transformations involved.[13] Group theory does not supply any "pictures" but only the structure of a set of operations: invariants of groups of transformations. Representing the invariance through pictures perhaps would be more intuitive, familiar, and pleasing to some than mathematically describing the structure of a set of operations, but to create such pictures one would inevitably have to connect the transformations with extraneous and irrelevant features, thus losing a degree of abstraction and generality.

Just as projective geometry studied how a series of mathematical objects in two dimensions are related as the projections of one (unprojected) mathematical object onto a plane via various transformations, so modern theoretical physics is partly concerned with studying how a certain series of physical objects (eg, particles presented in programmed performances in laboratory environments) can be viewed as different transformations of a single, fundamental object, which itself does not and cannot appear in laboratory environments "in toto" but only as one of several possible transformations.[14] The deeper symmetry or invariance is not visible in the single appearance of a particle (as it is not visible in the solitary ellipse), but only when other particles are considered. Invariances are exhibited only under multiple performances—multiple transformations of the system.

> *In the early 1960s, physicist Murray Gell-Mann was attempting to develop a new classification scheme for elementary particles, to replace the crude taxonomy by "weight classes": leptons, mesons, and baryons. Gell-Mann eventually hit on the idea of using group theory to create a classification scheme; certain particles, he saw, could be considered to belong together because they could be treated as the same particle "transformed" in various ways. The only problem was that not enough particles were known to fill out the groups Gell-Mann was considering. Nevertheless, he worked out the details of the scheme. Because a family of eight baryons fit the scheme most neatly, he called it, jokingly, the "Eightfold Way." In June of 1962, he attended a conference in Geneva in which the discovery of two new particles was announced. Those particles neatly fitted in to one of Gell-Mann's unfilled groups, a decimet, leaving only one "hole." Gell-Mann boldly announced that experimenters ought to look for a particle fitting that description—a particle that was essentially "the same" kind as those already known, but with different transformations—and he wrote down the pertinent information on a napkin and handed it to an experimenter. The particle, the "omega-minus," was discovered two years later.[15]*

A phenomenon may have to be transformed again and again with different kinds of transformations before what is invariant under all of them shows itself. Hence the preoccupation of scientists with "repetitions" of experiments, which are really multiple and different performances rather than repetitions inasmuch as they involve different performance contexts rather than exact replicas.

The widespread use of groups in physics is further facilitated by the noncommutativity of some rotation groups, which is to say that if I rotate a jagged rock a number of degrees in one direction and then a number of degrees in another direction, the profile I see is different than if I execute these operations in the reverse order. Certain quantum properties, as Dirac, Pauli, Heisenberg, and others discovered in the 1920s, are also noncommutative, and group theory offered a broad range of possibilities for representing such symmetries. As Steven Weinberg once remarked:

> Particles are bundles of energy and momentum. What are energy and momentum but the quantum numbers defined by [time and space] translations? What is angular momentum but the quantum number which is defined by rotation? In a sense, if you have an elementary particle, and you describe how it behaves under various symmetry transformations, including translations, rotations, gauge transformations, then you've said everything there is to say about the particle. The identity of the particle is fixed by its symmetry properties. The particle is nothing else but a representation of its symmetry group. The Universe is an enormous direct product of representations of symmetry groups. It's hard to say it any more strongly than that.[16]

Much of the progress of contemporary physics has gone hand-in-hand with the application of ever more sophisticated mathematical means of representing invariances to discover ever deeper symmetries. Weinberg has also described the excitement of physicists upon discovering new mathematical means of representing invariances, for this implies new possibilities for physicists of discovering symmetries and invariances in the physical world: "It's as if you were mining for some kind of metal—gold, say—and you'd done all the surface mining and you can't find any more, and then you find out that gold is also found underground in certain kinds of rock formations. You get excited not because it's deep underground, but because there's still some gold you haven't found yet."[17]

Groups have greatly expanded the variety of invariances in the world for physicists to discover and describe. One might say that a "thing" for a physicist is something with a space-time invariance; group theory thus has allowed physicists greater access to things of the physical world. Recalling our expanded conception of perception as well as the ancient meaning of *theorein* as the witnessing of something firsthand, we might say that group theory has allowed scientists to "see" ever deeper features of the world.

Theory, then, does not represent the object pictorially, but represents a transformation group that gives access to the object in all its variety. Even a model such as the double helix model of DNA has a group structure, for it describes a geometrical structure, and all geometrical structures are invariants in relation to spatial transformations of the object. The material components, too—the pairs of nitrogenous bases (adenine, cytosine, guanine, and thymine)—are also invariants, for they are not dependent on individual molecules, but only ones of a certain kind. There is thus a group theoretic replaceability by different individuals of the same kind. Wherever there is variation and invariants, group structures are present.

Mathematics, and indeed the entire theoretical vocabulary, is involved in two different semantics, one having to do with the description of the object, the other with the abstract model of the phenomenon. Consider the difference between the semantics involved in a musician referring to a note in performance and a music theorist referring to a note in the score. In the former case, the semantics are descriptive and the description can be fulfilled in many ways or even not fulfilled—one can anticipate one note but the performer will play another. In the latter case, the semantics are concerned with ideal objects, and consistency rather than fulfillment is at issue. For the laboratory experimenter, on the other hand, the language of "electrons" is descriptive, because it refers to things that show themselves in the equipment through measurement as fulfilling a certain description and that show themselves as having invariant space-time structures; measurements decide whether something is an electron rather than, say, a muon or positron. The language of the theorist, however, is not descriptive; the theorist's "electron" is a term in an abstract model, hence an ideal object. In this language, electrons are not (experienced, perceived) phenomena, but elements of an attempt to represent groups of transformations. Mathematics—that is, numbers—can be involved in both semantics; they can be part of the description of the object in the form of measurement, or of the representation of the invariance. Recognizing a finch may not involve numbers, but theories of finch demographics may. The representation of DNA as a double helix, or the representation encapsulated in the statement that "DNA makes RNA makes protein" may lack mathematics, but the laboratory notebooks that led to these expressions are full of mathematics. Or, mathematics may be involved in both experimental description and theoretical representation. In electromagnetism, for instance, numbers may be involved both in the descriptions of the objects, electrons, as well as in the theories, such as Maxwell's equations or quantum electrodynamics, used to represent the invariance behind these profiles. In the one case, the numbers are attached to the objects through descriptions of their profiles, while in the other case the numbers are attached not to the object but to the representation of the invariant of the profiles. In the former case, the effectiveness of the use of mathematics in the natural sciences is no greater mystery than in its use in carpentry. In the latter case, the effectiveness of mathematics in the natural sciences is no greater mystery than the reason why notes should represent music, or watches should be effective in dealing with train schedules.

The theories of physics, however, are only one kind; other branches of science involve other kinds of theories, including theories of behavior, of evolution, of geological change, and so forth, that do not rely so heavily on mathematics in the expression of invariances. Yet this does not make such theories any less scientific, and we would have every reason to suspect a philosophical account of scientific practice that took as its paradigmatic example of a theory one mathematically based rather than a theory of, say, geology or psychology, and then had to struggle to account for the legitimacy of these other kinds of theories. In the perspective being developed here, that problem does not arise. For, in this account, the principal aim of science is to present, represent, and recognize phenomena, and while mathematics can assist in the execution of these activities, each of them can also be executed without it.

Path-Dependency: Classical versus Nonclassical Phenomena

In 1935, Austrian physicist Erwin Schrödinger invented a "diabolical device" to illustrate the bizarre implications of quantum mechanics, the theory of the subatomic domain that he had helped to establish a decade earlier. Imagine, Schrödinger said, that a cat is sealed in a steel box, along with a Geiger counter, a small amount of radioactive material, a hammer, and a vial of hydrocyanic acid. The device is arranged so that when an atom of the radioactive substance decays, the Geiger counter discharges and, through a triggering mechanism, causes the hammer to smash the vial and release the poison. The problem is to describe the condition of the cat after one hour has passed—without looking inside the box.[18]

Common sense, of course, decrees that an atom will or will not have decayed, and the cat will be either dead or alive but not both. Radioactive decays, however, take place according to the laws of quantum mechanics, where things are different. According to the standard interpretation of the theory, such decays are indeterminate—they neither take place, nor don't take place—until someone performs a measurement, which in this case means opening the box and examining the cat. Until then, Schrödinger wrote, "the living and dead cat are, pardon the expression, blended or smeared out."[19]

Schrödinger thereby gave graphic articulation to a problem that has plagued the philosophy of science ever since, concerning the "meaning" of quantum mechanics for the understanding of the world. The trouble stemmed largely from the role of what is known as the Schrödinger wave equation, invented by Schrödinger himself in 1926, which depicted much of what transpired inside an atom in terms of waves. The wave function for a particle consists of a superposition of possible states of the particle. When a physicist performs a measurement, the wave function "collapses"; all but one possibility is excluded, and the experimenter winds up with actual values.

Common sense is not faced with a problem if, as Einstein argued, quantum mechanics is incomplete, and the wave equation simply indicates the presence of other factors, or "hidden variables," that, when discovered, will allow scientists to describe the behavior of subatomic particles as exactly as that of marbles. The wave equation then represents an ensemble of systems, and we are simply ignorant of what factors determine which system will prevail.

But, as Niels Bohr and other participants in the Copenhagen circle (such as Heisenberg) argued, quantum mechanics is complete; one cannot know more than what the Schrödinger wave equation is able to say. Until an experimenter measures the system—an electron, say—and the wave function collapses, it does not have a definite position, momentum, energy, and so on. The Copenhagen group appreciated that the implications of quantum theory reverberated far beyond physics to challenge traditional ontology as well, and in particular the relation between theory and the world. Quantum mechanics signified that human beings were simply not equipped to visualize the totally different world "down there." Whereas classical physics depicts objects as having properties all of which could be specified in principle at any given time, quantum mechanics requires us to regard some of these properties as "complementary but exclusive." That is, some property can be specified precisely at one particular time only at the cost of precision in specifying another property, the choice being left up to the experimenter.

This idea was the cornerstone of what was soon called the Copenhagen Interpretation, which is still the most important elaboration of the meaning of quantum mechanics. For Bohr and others, the Copenhagen Interpretation was a kind of philosophical key allowing them to feel comfortable with the peculiarities of the subatomic realm. Nevertheless, it forced physicists to face a bewildering situation, for it seemed to say that conventional assumptions about the nature of reality do not apply to the subatomic world. The Schrödinger equation, Bohr said, does not depict actual, independent objects: "[A]n independent reality in the ordinary physical sense can neither be ascribed to the phenomena nor to the agencies of observations."[20] This is the problem beautifully illustrated by Schrödinger's Cat Paradox.

Philosophers of science frequently attempt to address this issue *epistemologically*, in terms of the structure of the object known.[21] The problem, however, is that the classical epistemology appealed to in such discussions, based on classical physics and involving a pregiven space and time occupied by objects in a continuous space-time trajectory, is inadequate for treating quantum phenomena, which have to do with phenomena of a different kind. To address this problem properly, one has to do so *ontologically*, in terms of the structure of the knowing act. Failure to do so has given rise to misunderstandings that include assertion of the existence of an important connection between quantum mechanics and Eastern mysticism.

Quantum mysticism, as this view might be called, had its origins in some statements by certain of the pioneers of quantum mechanics and blossomed in the 1970s and 1980s. The groundwork for it was laid by Copenhagen scientists themselves; when Niels Bohr was knighted in 1947, for instance, he selected for his coat of arms design a yin-yang symbol, while Heisenberg and Robert Oppenheimer, among others, sometimes drew connections between their work and ideas of Eastern religions. The founding father of the modern movement, however, is Fritjof Capra, a maverick physicist whose *The Tao of Physics* sold millions of copies as it asserted a connection between physicists and Zen Buddhism; indeed, writes Capra, this connection is of critical significance for it demonstrates the inadequacy of our present world view and points to the need for a change so drastic as to amount to cultural revolution: "The survival of our whole civilization may depend on whether we can bring about such a change."[22] A nonscientist, Gary Zukav, then produced *The Dancing Wu Li Masters*, which made the breezy announcement that "philosophically . . . the implications [of quantum mechanics] are psychedelic," and won an American Book Award.[23] Still more visible and effusive has been actress Shirley MacLaine, who quotes Zukav in the epigraph of her book *Dancing in the Light*.[24] Things have got a little unfastened in the world of scientific popularization, and quantum mechanics is the culprit.

Quantum mysticism was created by the awareness (shared, as we saw in chapter 2, by Dewey), that quantum mechanics forces us to reject the traditional, "mythic" account of science. I, too, have argued that the account must be rejected, but without arguing that Eastern mysticism offers a satisfactory alternative. The problems with quantum mysticism cannot be exhaustively detailed here, and I shall only sketch them out.[25] Sal Restivo has critically examined quantum mysticism (his word for it is *parallelism*), in *The Social Relations of Physics, Mysticism, and Mathematics*. He observes:

> The basic data for parallelism are common language (for example, English) statements on the nature and implications of physics and mysticism that vary in technical content. The methodology of parallelism is the comparative analysis of such statements. Similar rhetoric, imagery, and metaphoric content in such statements constitute the evidence for parallelism. The basic assumption in this approach is that if the imagery and the rhetorical and metaphoric content of statements on physics and mysticism are similar, the conceptual content must be similar, and the experience of reality must also be similar among physicists and mystics.[26]

But even before such comparisons are made, Restivo points out, the success of such a procedure must first depend on proper selection and translation of representative texts from both mystics and physicists. Several perils stand in the way. The first lies in selection—the texts chosen must truly reflect the spirit of Eastern mysticism and physics and neither be idiosyncratic nor overlook conflict and diversity. Restivo finds, however, that Capra and others tend to pick and choose their texts, ignoring the immense variety in mystical experiences and assuming that the world views of Buddhism, Hinduism, and Taoism are essentially the same. It might be added that quantum mystics also tend to pick and choose their physics texts as well, sweeping under the rug arguments about the proper nature and direction of the field. Here we can cite one of Capra's own examples, the bootstrap hypothesis, which denies that there are fundamental particles and asserts that each type of particle is linked with all the others. In the mid-1970s, when Capra's book was written, this model was popular among some scientists. Others, however, embraced the quark model, based on the distinctly unmystical notion of fundamental particles. By promoting the bootstrap hypothesis as typical of the world view of modern physics, and then comparing it to certain Eastern modes of thought, Capra uses an idiosyncratic picture of physics to buttress his quantum mystical claims.

Another peril in the quantum mystical methodology is contamination. Many concepts of modern physics have filtered into common vocabulary, including *space-time, complementarity, quantum, relativity,* and so forth, as have many concepts of Eastern religions. This may lead to conscious or unconscious misuse of terminology, suggesting links where there are none. An example of deliberate, playful misuse is Gell-Mann's 1962 system of representing strongly interacting subatomic particles that assigned the baryons to one group of eight; Gell-Mann named his system the "Eightfold Way" in joking homage to the Buddha's teaching. Any resemblance to actual teachings of the Buddha is, of course, superficial.

But far and away the most formidable peril is that of translation. A prerequisite for comparing statements by physicists and mystics is that both be in a similar language. What must be translated, however, is in each case profoundly different. The theories of modern physics, for instance, are couched in a mathematical formalism that requires years of training to understand and are inevitably rendered inadequately when put in non-mathematical language; Feynman once claimed that it is impossible to explain the meaning of the laws of nature to those unfamiliar with mathematics.[27] Language may therefore have entirely different relations to what is addressed in the hands of physicists and mystics. If so, both the conceptual content of what physicists and mystics say and their experiences of reality are likely to be different even when they speak in similar language. Zukav's assertions that "physics is not mathematics" and that most physical ideas are "essentially simple" do not reassure; neither does Capra's statement regarding Hinduism, Buddhism, and Taoism that "the basic features of their world view are the same."[28]

Eastern mysticism, moreover, is less like a body of truths than a program for a spiritual quest, an attitude adopted toward the world, the gods, the transcendental. The wisdom expressed by this attitude, according to its practitioners, has remained essentially unchanged for thousands of years. Science, however, continually evolves, and even its fundamental concepts are subject to change. If quantum mystics argue that contemporary scientists are making the *same* claims about the world that ancient Eastern mystics make, is ancient Eastern mysticism refuted when and if new discoveries force scientists to revise their views? Capra's bootstrap hypothesis is again a good example. In the late 1970s, after Capra's book was published, the quark model triumphed decisively over the bootstrap model. This triumph was wholly traditional; quarks explained (scripted) the data better. Must Buddhists now cast their world view out the window?

Quantum mystics wind up distorting both science and Eastern mysticism. In general, they distort science by deemphasizing the role of mathematics, mathematical reasoning, and technology, claiming that what is essential about science can be cleanly separated from these. They distort Eastern mysticism by overlooking the social and historical roots of the various Eastern religions and the extent to which the cosmologies of Buddhism, Hinduism, and Taoism not only were individually shaped by the structures of the societies that gave birth to them but also evolved in the light of exposure to changing social conditions. The convenient neglect by quantum mystics of such all-important elements of Eastern religions as rejection of the world, the long period of discipleship, and the preoccupation with death indicate that today's "quantum Zen" is probably as ignorant of the real religions of the East as was the "California Zen" of a generation ago.

This digression into quantum mysticism is worth pursuing because today quantum mysticism appears to be on the verge of becoming as firmly entrenched in popular culture as astrology. Like advocates of creationism and astrology, quantum mystics misrepresent science to the public, and the result is detrimental to the field. Moreover, conceptual confusion in physics, as in any area, is worth trying to clear up for its own sake.

The schema I have proposed, which sees experimentation as involving the making-appear of scripted phenomena through profiles, reorganizes the issues that led to quantum mysticism in a way that makes it an unnecessary response to them. Certain phenomena can be manipulated so that every one of their profiles can in principle be sampled, with the sampling of one set posing no bar to the later sampling of any other set. If I turn over a crystal, for instance, I can examine each of its facets in turn, and even though at each point others are hidden from view, I can recover them by further turns of the crystal. Let us call phenomena with this characteristic *classical phenomena*. Certain other kinds of phenomena, however, cannot be manipulated in this way, for the appearance of certain profiles entails the loss of the ability of certain other profiles to appear. Let us call phenomena with this characteristic *nonclassical phenomena*. Nonclassical phenomena have the property that their data are *path dependent*—the order of the decisions that take place in the environment affecting the appearance of the phenomena affects the kinds of profiles not only that it has at any given moment but that it can have in the future.[29] These environmental decisions may or may not have been made by human beings.

Quantum phenomena, for example, are nonclassical phenomena where the decisions are of human origin. The dependence of the sets of data or profiles of quantum phenomena on the noncommutative order of a series of preparation decisions made in the laboratory concerning certain pairs of terms (momentum and position, energy and time, etc.) is well known and need not be recounted in detail here:

> The elements of each pair, though mutually exclusive in their historical realizations, belong nevertheless to the definition of the same phenomenon insofar as this is full of real historical potentiality. We could speak of the same phenomenon as "fleshed out" by and for a local research community by the sequence of decisions made by that community; it is this sequence that determines the historical path of its evolution in the local world of that local community. The phenomenon remains, however, throughout its evolution the *identical* phenomenon.[30]

An example of a nonclassical phenomenon where the "decisions" are of nonhuman origin can be found in Steven Jay Gould's book, *Wonderful Life: The Burgess Shale and the Nature of History*, which challenges the iconography of evolution as either a ladder of progress or a cone of increasing diversity. Instead, it is a "copiously branching bush, continually pruned by the grim reaper of extinction." The pruning process is carried out by a particular sequence of environmental decisions, and other forms would have arisen if another sequence had been followed. Gould dramatizes this by carrying out a thought experiment which he calls "replaying life's tape."[31] "You press the rewind button and, making sure you thoroughly erase everything that actually happened, go back to any time and place in the past—say, to the seas of the Burgess Shale. Then let the tape run again and see if the repetition looked at all like the original."[32] Were evolution a ladder of progress or a cone of increasing diversity, approximately the same forms would emerge. Gould, however, claims that radical decimation created by environmental decisions is an overriding factor in the development of life forms, leading him to propose a "third alternative":

> I believe that the reconstructed Burgess fauna, interpreted by the theme of replaying life's tape, offers powerful support for this different view of life: any replay of the tape would lead evolution down a pathway radically different from the road actually taken. But the consequent differences in outcome do not imply that evolution is senseless, and without meaningful pattern; the divergent route of the replay would be just as interpretable, just as explainable *after* the fact, as the actual road. But the diversity of possible itineraries does demonstrate that eventual results cannot be predicted at the outset. Each step proceeds for cause, but no finale can be specified at the start, and none would ever occur a second time in the same way, because

any pathway proceeds through thousands of improbable stages. Alter any early event, ever so slightly and without apparent importance at the time, and evolution cascades into a radically different channel. This third alternative represents no more nor less than the essence of history. Its name is contingency—and contingency is a thing unto itself, not the titration of determinism by randomness. Science has been slow to admit the different explanatory world of history into its domain—and our interpretations have been impoverished by this omission. Science has also tended to denigrate history, when forced to a confrontation, by regarding any invocation of contingency as less elegant or less meaningful than explanations based directly on timeless "laws of nature."[33]

In our terms, the evolution of life as Gould describes it is a path-dependent nonclassical phenomenon. The path dependency of this kind of phenomenon does not mean that its study is any less scientific than that of classical phenomena, though it does entail a different kind of "scripting." Narrative, or the organization of information into roughly sequential order exhibiting the decisions affecting the phenomenon, may need to figure in describing how profiles are related, and thus the nature of the phenomenon involved. Or, one may choose to characterize the invariant by focusing on an exemplary incident or object; this is precisely how Gould describes the aim of his book, which centers on "the new interpretation of the Burgess Shale as our finest illustration of what contingency implies in our quest to understand the evolution of life."[34]

Theory is thus neither strategy, instrument, nor eternal description of entities above time and history. It is not a strategy, for it is an activity of a fundamentally different kind than that of bringing something into being, of producing or realizing performances. Theorists participate in the performance event as observers, experimenters as producers. Consider how out of place a theorist would be attempting to become involved in lab work, or a music theorist or audience attempting to become involved in the details of performance.[35] Of course, some remarkable individuals have performed skillfully both the functions of representer and performer (Fermi and Shakespeare come to mind) but this testifies to their ability to become fluent in two different functions. A theory is not an instrument because it is more intimately related to the incarnated presence one wishes to obtain than a tool to the object that the instrument is used to create. A theory is not merely an agent (like a musical instrument) that one uses to create something; it programs the use of that instrument, it tells one how to use the instrument that one uses. Finally, a theory is not an eternal description because the praxes used to get from representation to represented are not eternal but belong in human time and history.[36]

Wigner's mistake, and that of many others working within conventional perspectives, is to view the relation between mathematics and the world in parallelist terms—that there are physical invariants and mathematical invariants, and the two have the property of mirroring one another in a way that can only be described as uncanny. But even though mathematics may be treated as an independent symbol system, in its use in theories of the physical world its relation to the world is not one of mirroring but of process and product. The mistake of Heidegger and many other phenomenologists, on the other hand, is likewise to regard theories as independent sets of abstract symbols, but to assume that these symbols seek to describe something beyond the involvement of human beings with the life-world. The problem with many phenomenological accounts of natural science is that they do not make use of a sufficiently expanded conception of perception. They have had a tendency to regard natural science with childish naïveté, as if a phenomenology of experimentation could be carried out by describing what a child or scientific illiterate would see if brought into the experimental situation. The experimenter, however, is no child, but a trained individual whose perceptual ability has been enriched by readable technologies and who is on the lookout for invariances in an environment enriched by those same technologies.

The conceptual problems that can emerge when theory is taken as structure rather than as scripting are evident in the confusions surrounding the *meaning* of various aspects of quantum mechanics, such as those that give rise to quantum mysticism, which concern especially the reduction of the superposition of states and the nonlocality seemingly required by Bell's theorem. Traditionally, objectivity has been regarded as belonging to those things whose fundamental characteristics can all be present, at least in principle, at the same time, with the guiding image being that of the thing's presence to a divine knower. This view corresponds to the theological one expressed in the image of the book of Nature, where a divine blueprint is already written down. But as I have already mentioned, the value of this image has been challenged in this century by the introduction of mathematical forms into theories that rendered ambiguous the way theories speak about the world. Quantum mechanical theory cannot be thought of as picturing states of real phenomena independently of specific measurement contexts. The simplest way of expressing the reason why is that quantum phenomena are represented by infinite-dimensional ket-vectors in Hilbert space, precluding the possibility of a complete "picture" of a quantum state.

The solution of the Copenhagen circle, however, amounted essentially to intellectual capitulation, to conceding that this relation was unintelligible, a mystery. Quantum theory was to a large extent to be thought of merely as a calculational procedure, and one can and should ask no more of it. One cannot use it to shed light on the actual state of the world, which is why Frank Wilczek, for instance, described the Copenhagen interpretation as one of "renunciation."[37] Philosophers of science have sometimes attempted to explain these mysteries by approaching them as purely epistemological problems, with less than satisfactory results.

The questions at issue here are both important and difficult. Nobody, says historian of science Stephen Brush, "has yet formulated a consistent world view that incorporates the CI [Copenhagen Interpretation] of QM [Quantum Mechanics] while excluding what most scientists would call pseudo-sciences—astrology, parapsychology, creationism, Velikovsky's theories, and thousands of other cults and doctrines."[38] The reason ultimately motivating quantum mysticism might be expressed as follows. The traditional Western view allegedly undermined by quantum mechanics involves the notion that objectivity belongs to those things whose fundamental characteristics can all be present at the same time, at least in principle. The guiding image here is of a divine mind capable of witnessing the spectacle of the world's *eidoi* at a glance; the things of the world must all be simultaneously intuitable to the Demiurge. The ultimate aim of scientific theory, in this traditional view, is to develop a similarly complete picture of the "fundamental furniture of the world." But quantum mechanics reveals the impossibility of this aim. Some properties of certain quantum objects are fleshed out differently in different contexts, those contexts depending, in the laboratory, on the actions of the experimenter. Quantum mystics then argue essentially as follows: If theory aims to picture the basic things of the world, and if modern physics shows that such a complete picture is not possible and that every picture is finite and only emerges when we engage with the world, then the world as it emerges in our theories is an illusion, and reality is our own creation. Put this way, the discoveries of quantum mechanics do indeed suggest some similarities to Eastern mysticism, at least as they are popularly and cursorily understood in the West.

But the failure of traditional science to elucidate the *meaning* of quantum mechanics does not mean that the field should be ceded to the Eastern mystics. The solution resorted to by the Copenhagen circle—to sever entirely the ontological connection between the abstract image provided by theory and the world—hints that the difficulty lies (as quantum mystics suggest) in the initial assumption that theory addresses something whose meaning is perspicuous independently of the world. But attempts to resolve the difficulty by appealing to likenesses with the supposed worldview of Eastern mysticism, as I suggested, founder on the immense problems involved in forging the alleged analogy between mysticism and modern physics.

The beginnings of a better solution can be found by looking once again to the dramatic arts, whose practitioners have long known that the meaning of lines of dialogue depends to a great extent on how the work is performed, and that ambiguities in the meaning of passages in the script cannot all be textually decidable. There are even cases where different performances of an identical passage of text can have diametrically opposed meanings, as the often spectacularly diverse productions of *Hamlet* suggest (e.g., his various claims to sanity or insanity). And when Petruccio says, "Kiss me, Kate" at the end of *Taming of the Shrew*, the line is frequently taken as the final expression of Kate's humiliation and defeat. In a recent Royal Shakespeare Company production, however, the line instead denoted her triumph and his shame and defeat in his attempts to break her. After Petruccio's request, she bent, all in irony, to kiss his foot; he withdraws in horror at what he has tried to do to her. Let us not say that the ability of this line to have opposite meanings is due to ambiguities or inadequacies in the direction, to the incompleteness of the play, or to the incompetence of the playwright. We must refer it to the principle of the primacy of performance: that the meaning of a play is fully realized only in performance, and that the range of meaning possibilities can be quite broad in certain cases depending on the set of decisions involved in the production.

In the case of quantum mechanics, such considerations suggest that the conceptual problems concerning the nature of quantum mechanics must not be approached from the perspective of epistemology, as have traditional philosophers of science, but from the perspective of the ontology of the experimental production of knowledge. Within this perspective, these conceptual problems can be clarified by considering quantum phenomena as an instance of the path dependence of nonclassical phenomena. The path dependence of quantum phenomena then are no more mysterious than the path dependence of the forms of animal life as suggested by Gould's thought experiment "replaying life's tape." The same phenomenon is involved, even though its profiles are dependent in a noncommutative way on the path of decisions.[39]

Quantum mechanics does indeed pose a serious challenge to traditional philosophy of science and its notions of objectivity. But this does not necessarily mean that the tenets of Eastern mysticism meet that challenge, or even that the tenets of mysticism are invalidated if they fail the challenge. It does mean that the philosophy of science will have to work out its notions of theory, objectivity, and scientific phenomena in more detail—which can be done, I think, with the aid of the schema I have suggested. Until then, Schrödinger's spectral feline will continue to haunt quantum mechanics with occult implications. The misguided search for mystical insight can be taken as an indication of how much work physicists and philosophers still have before them.

VI

PERFORMANCE
RECOGNITION

The perception of a new phenomenon in nature is generally couched in the language of *discovery,* as the coming across and identification of fully constituted phenomena (albeit whose properties may be yet undiscovered) by an individual or individuals. What has been said about perception in previous chapters should suffice to indicate the inadequacies of this language, for to perceive something as a thing is simultaneously to come to apprehend invariants of some variety, that is, to have constituted it. To perceive is to perceive something with structure; it is to have anticipations of the way it appears in other contexts, whether these anticipations are eventually fulfilled or not.

The existence of inadequacies in the language of discovery has been known for some time.[1] To expose some of them, let us consider one of the classic instances of scientific discovery, that of X rays. No original historical research is required, only a recital of well-known information.[2]

On the afternoon of Friday November 8, 1895, German physicist Wilhelm Conrad Röntgen was working alone with a cathode ray tube in his Würzburg laboratory when he noticed that a fluorescent screen on the same table as his apparatus had commenced to glow, for no reason that he could imagine. He turned off the switch, and the glow ceased. He turned it on again, and the glow resumed. Quite a simple presentation, with a minimum of equipment, skill, and planning—but utterly baffling. He was soon able to determine that the cause of the glow came from something emitted at one end of the cathode ray tube; that this something was able to penetrate paper, wood, human flesh, and even thin pieces of metal; and that it did not behave at all like the cathode rays he had been examining. Nevertheless, Röntgen remained so astonished that, as he told his wife later, he had to do these procedures over and over again to be sure that he was not seeing things. After eight weeks of experiments, he was finally able to convince himself that the effect was real. Not that he had any idea of what it was. Indeed, when on December 28 he mailed out a "Preliminary Communication" entitled "On a New Kind of Rays," he underscored their still-cryptic nature by the name he gave them, "X Rays." Colleagues all over Europe were stunned. Every well-equipped lab had the equipment to produce the rays, and it was inconceivable that no one had seen such rays before. Small wonder that the initial announcement of these rays and their astonishing powers provoked skepticism. So eminent a scientist as Lord Kelvin pronounced them a hoax, while one of Röntgen's friends called his account a "fairy tale." Within a week, however, physicists reported producing the rays, with results similar to Röntgen's. Later, other scientists recalled noticing strangely glowing screens in the vicinity of cathode ray tubes, but had not realized their significance.[3]

Thus transpired one of the simplest, most clear-cut, and most momentous cases of a discovery in science. The 15th edition of the *Encyclopaedia Britannica*, published in 1990, states the matter straightforwardly: "X-Rays were discovered in Würzburg, Germany on November 8, 1895, by Wilhelm Conrad Röntgen."[4] But even in this most transparent of discovery tales, as Thomas Kuhn has noted in *The Structure of Scientific Revolutions*, the language of discovery conceals much about what actually took place in the events leading up to the "Preliminary Communication." Complexities and ambiguities appear as soon as one begins to explore two apparently easy questions: Who discovered X rays? And when was the discovery made?

The answer to the first question seems obvious enough: Röntgen. When the Nobel Prizes were established in 1901, he received the first physics prize for the discovery of X rays. But the question is not so simple. As in the case of many a scientific discovery, no sooner had attention been drawn to Röntgen's accomplishment than a number of other claimants for the title of discoverer surfaced. Most claims were fraudulent, but a few were not easily dismissable. In 1879, for example, British physicist Sir William Crookes noticed that photographic plates in the vicinity of his cathode ray tubes became fogged, and he returned several to the manufacturer as defective; he had seen X rays, some argued. Other experimenters, including Eugen Goldstein and Nobel laureate Philipp Lenard, had observed the inexplicable flourescence of certain compounds in the presence of cathode ray tubes; for a time, Lenard was promoted by a certain community of British scientists as the true discoverer of X rays. But the most uncanny episode in the "prehistory" of X rays transpired in the laboratory of A. W. Goodspeed at the University of Pennsylvania, in Philadelphia:

On February 22, 1890, Goodspeed and a friend spent some time photographing electric sparks and discharges. After they had finished, they brought out a cathode ray tube and tinkered with it for some time. When they developed the photographic plates later, they were astonished to discover the image of two strange discs on one of them. They could not explain the presence of this image until six years later, when news of Röntgen's work prompted Goodspeed to dig out the photograph and the cathode ray tube, recreate the initial conditions, and make a new photographic exposure. Goodspeed concluded that his original plate had been exposed by X rays. He never claimed to have discovered the rays. But two months after Röntgen's announcement, Goodspeed did claim—bizarrely but not unfairly—to have been the first photographer of X rays. "All we ask is that you remember, gentlemen, that six years ago, day for day, the first picture in the world by cathodic rays was taken in the Physical Laboratory of the University of Pennsylvania."[5]

As these stories show, discoveries are not made by the first person who merely encounters or even engages with what is subsequently perceived to be a manifestation or profile of the new phenomenon. If that were the case, the researchers who had noticed strangely fluorescing screens or inexplicably fogged photographic plates would have been accorded priority for the discovery of X rays. Goodspeed had actually taken X ray photographs—yet when Röntgen took his six years later, he received the Nobel Prize. Nor does discovery necessarily involve the formation and confirmation of a hypothesis. Röntgen had no hypothesis about X rays before he observed them, whence his utter astonishment. The hypothesis he did offer, ever so tentatively, at the end of his "Preliminary Communication," turned out to be, as he suspected at the time, completely wrong. By what right, then, did Röntgen earn his title as "discoverer"?

A similar ambiguity surrounds the second of the two "easy" questions, regarding when the discovery of X rays took place. The answer confidently supplied by the *Britannica* is November 8, when Röntgen first noticed the glow. Obviously, the discovery was not made before November 8 or after December 28. Between those two dates, Röntgen had to repeat his work many times to convince himself that it was real. For some period of time he did not know whether he could believe his eyes and had no idea what kind of thing—hallucination or real phenomenon—he was experiencing. We must imagine him slowly acquiring the confidence that he was in the presence of a real phenomenon, and we must consider that achievement of confidence in the bodily presence of something in the world—whenever that achievement was, and regardless of whether it took place in an instant or over time—to be the decisive moment of discovery.

What characterizes that confidence, and how is it acquired? Röntgen was not confident of the reality of the phenomenon when he first encountered it. Nor was his confidence the outcome of a process of hypothesis formation and confirmation. Rather, that confidence came from Röntgen's apprehension of an *invariant*. Röntgen convinced himself that he was seeing in the images on the photographic plates and the glowing screens different sides of the *same* phenomenon, something that behaved in certain ways under certain circumstances. By apprehending the existence of a previously unknown invariant, he came to experience thereby the bodily presence of something in his laboratory—the fact that the phenomenon *existed*. With that confidence, the experiment—and the discovery—was complete.

To be sure, there are many different kinds and circumstances of discovery, and many different kinds of entities discovered. But even one elementary discovery story can suffice to raise questions about the nature of the discovery process. One may try to view discovery as a problem of psychology, and attempt to describe the psychological conditions under which discovery takes place.[6] Another tack is to try to view the issue in terms of the creation and justification of hypothesis; it has been argued that there is a "logic of discovery" having to do with the reasons for offering a certain hypothesis, and that there is no such logic of discovery different from the logic of hypothesis justification.[7]

Discovery as Recognition

If scientific entities are viewed as analogous to perceptual objects, with the difference that scientific entities are perceived through readable technologies, presented in experimentation, and represented through theories, then the name for the aforementioned achievement of confidence is *recognition*. Recognition is the perceptual apprehension of a phenomenon *as* a phenomenon when what is at issue is its bodily presence. If I momentarily leave a room of a dozen or so familiar faces—the table at dinner time, for instance—and then return a moment later, I do not say that I have so many near-instantaneous acts of recognition of these individuals; I "know who they are." Rather, I recognize people when I come across them in an environment where their presence is an issue for me. That "presence" is a concern; I *linger over* or am *arrested by* it. Searching through faces in a chorus, I am delighted finally to recognize my friend Fred; in a sea of faces on the dance floor, I cringe when I recognize my ex-friend Judi. In recognition, I experience a bodily presence that amounts to the achievement of a new perception, though "new" here does not necessarily mean of an object with which I am unfamiliar.

Basic research is the activity that aims to recognize new phenomena in the world and their properties, and experimentation is ordered toward such recognition. One might distinguish further between aiming toward the recognition of a suspected new phenomenon (Röntgen's research when he was puzzled but before the achievement of confidence), and aiming toward recognition of features of an already recognized phenomenon (Röntgen's research thereafter). The recognition could be of an altogether new phenomenon (Röntgen's initial recognition), or of a phenomenon whose presence is already familiar and identified (that of his colleagues who reënacted his experiments). It goes without saying that one can encounter a phenomenon without recognizing it; what I have called the "prehistory" of a discovery involves engagement with a phenomenon without an explicit act of recognition. Röntgen's confidence grows as he comes to recognize that the same phenomenon is appearing in different manifestations; that he is seeing the same object, a mysterious form of radiation, in different circumstances.

The achievement of confidence is thus tied to recognition of the presence of a phenomenon. It makes no difference whether the recognition is made via readable technologies or the naked eye, for even when one perceives something via a readable technology, one needs to be sure that one is seeing a profile of a real phenomenon, not just an "artifact of the machine." It also makes no difference that the profiles of the phenomenon in question are so varied in appearance, from glowing screens to images of bone on photographic emulsion. Indeed, a hallmark of the movement of many sciences is to be able to see a myriad of different kinds of phenomena as but different profiles of one—as, say, when biology replaced the myriad aspects of phenotypic inheritance with the law of genotypic inheritance.

This centrality of the role of phenomena has implications for the "unity of science." Positivist-inspired philosophy of science viewed method as the key to this unity: "The only necessary unity is that of Method," writes Comte.[8] A similar approach was taken by the "Unity of Science" movement of the late 1930s, led by members of the Vienna Circle. While previous attempts to unify the sciences were based on metaphysical and theological principles, partisans of this movement saw them as unified by virtue of a common logic. "When we ask whether there is a unity in science," wrote Carnap, "we mean this as a question of logic, concerning the logical relationships between the terms and the laws of the various branches of science."[9] But the primacy of phenomena in science suggests that what unifies a science has to do instead with the unity of the kind of phenomena that it pursues; moreover, that the relations between the various branches of science have less to do with generality and reducibility than with relations between these different kinds of phenomena. The primacy of the phenomenon means that the phenomena themselves determine the method used to present, represent, and recognize them.

Recognition of a phenomenon, I said, involves apprehension of invariance, which in turn involves the presence of an anticipation of other profiles, and enough experience with them to be confident that they fulfill the expectations. In recognizing a new perception, one *constitutes* it—though as Röntgen's experience shows this constitution does not have to be a modeled or theoretical process.

But how can one recognize the truly novel? How does one set out to find something when one does not even know what it is one is looking for? And supposing one does find it—how could one possibly know that this was the very thing one was looking for? The problem is famously formulated in the Platonic Dialogue *Meno*, when Meno, frustrated, blurts out to Socrates, "And how do you inquire into that which you do not know? What will you put forward as the subject of the inquiry? And if you find what you want, how will you ever know that this is the thing which you did not know?"[10] Plato attempted to resolve the problem through his doctrine of recollection *(anamnesis)*. Even when we do not misunderstand this doctrine by thinking that Plato meant that we had previous incarnations in which we human beings knew things in the same way we do now, this response is unsatisfactory. What interests us about science is its capacity to add genuinely new features to the life-world; to be world-building rather than world-recollecting.

Aristotle on Recognition

In view of the importance of recognition, it is worth reviewing existing philosophical perspectives to provide a basic framework for discussing the issue. The problem of how science manages to recognize the new may be broached through explication of the concept of recognition, with the aid of a philosophical literature that reaches back to Aristotle. While it may seem odd to turn to Aristotle, of all people, and to his *Poetics*, of all places, for assistance in clarifying the paradoxes of the discovery process in science, this reference is made necessary by lack of appropriate treatment of the issue by traditional philosophy of science.

In Aristotle's *Poetics*, recognition *(anagnorisis)* is presented as one of the three essential parts of plot (which is in turn "the first essential and soul of a tragedy") besides reversal and suffering. Given the context, Aristotle naturally restricts his discussion to recognition between persons, although he accepts that one can recognize objects, events, and rules.[11] Even though Aristotle's focus is narrow, four features of recognition relevant to this discussion can be identified, either from his explicit account or his examples (of which the paradigm case is Oedipus). First, recognition is a transition from ignorance to knowledge; second, it consists of a perceptual act; third, it is the result of a concernful engagement with the world; and fourth, it has unanticipated consequences. These four features form a basis for developing a suitable account of recognition, which I shall elaborate when necessary with reference to other philosophical accounts, such as Hegel's (whose account of recognition is indebted to Aristotle's).[12]

First, Aristotle defines recognition as a *passage from ignorance to knowledge*. What becomes known in Aristotle's tragic recognition is the (unsuspected) identity of someone with whom one is already familiar, usually intimately. This recognition can come as a complete surprise, as when Creusa recognizes the identify of her long lost son by a casket of swaddling clothes, or as the fulfillment of one's (worst) suspicions, as the recognition that Oedipus achieves through his questioning of Tiresias, the Corinthian messenger, and Jocasta, that he himself is the slayer of Laius. In such cases, which come as the relaxation or resolution of dramatic tension, the moment of recognition is at the same time a moment of self-knowledge. Recognition of the other is at the same time self-recognition; it is the achievement of an understanding about one's real identity and place in the family or community. Aspects of oneself are disclosed, false ideas about oneself are cast aside, and the tension between oneself and the world is resolved. Hegel likewise focuses on the social aspect of recognition, accepting four senses of the concept: recognition as occurring between an individual and God, society, other individuals, and oneself.[13] Urmson, Sayre, Price, and others discuss explicitly the case of recognition of objects. Urmson distinguishes between recognition of individuals as of a new kind, of individuals as a member of an existing kind, and of a property of individuals; Casey speaks of recognition of styles and situations; Sayre discusses pattern recognition. In the scientific context, recognition involves apprehension of an invariant. Acts of recognition are made by both experimenters and theorists. Experimenters recognize the presence of an invariance through directly manipulating the machine conditions, varying the profiles themselves (which can take the routine form of varying the systematics or more elaborate forms) in order to witness its behavior in different circumstances. What is recognized can be of familiar or unfamiliar form; a different kind of recognition is involved in recognizing a new galaxy, element, or subatomic particle than in recognizing what it is to be a galaxy, an element, a particle. Theorists recognize the presence of an invariant through looking at the entire performance program that constituted the presence of the phenomenon and comparing it with others; Gell-Mann's classification scheme called the "Eightfold Way," for instance, was discovered as an invariance amid already recognized particles, which culminated in the abstract possibility of experimental performances of new phenomena (e.g., the omega minus particle).[14]

Second, recognition for Aristotle generally consists of a *perceptual act*, one which can be prompted by a number of vastly different means. Aristotle classifies these into six, ranging from signs such as scars and necklaces to "the incidents themselves," but other classifications are clearly possible. It is true that one of Aristotle's six means of recognition is "by inference" rather than by perception, which he illustrates with a scene from Aeschylus's *Choephoroe* in which Electra concludes that her brother Orestes is in the vicinity based on a lock of hair of a color and texture similar to her own and footprints of a shape and size similar to hers. But one might either challenge the validity of reference to inference in such cases, for even an apparently sound inference may be questioned as insufficient for genuine recognition (as, for instance, Electra herself does in that scene). Or, one could consider that what is involved is not an inference based on pieces of evidence, but are rather signs which initiate a perceptual act of recognition (and in fact Electra does not finally admit the presence of Orestes until he reveals himself and shows her the robe she herself had woven for him).

Traditionally, many discoveries are treated not as recognitions but inductions or inferences. Consider two cases of the discovery of new elements: Sir Norman Lockyer's discovery of helium in the Sun by observing a new line in the solar spectrum, and the discovery of radium by Marie and Pierre Curie, who noticed the presence of a strongly radioactive source in pitchblende. Are these instances of inference or recognition? I would argue for the latter. Lockyer's discovery was not a matter of deciphering the meaning of an unfamiliar, abstract mark on a piece of paper. He knew that the mark in question was a spectral line; he knew what kind of thing made spectral lines and what kinds did not; he knew under which conditions spectral lines appear and under which conditions they do not appear. Similarly, the Curies' discovery required an entire body of knowledge of how chemical elements behaved using the techniques of analytical chemistry. What transpired to make both discoveries possible was much more than the ascent to a general law on the basis of a few pieces of data, but rather a familiarity with an entire system of closely linked behaviors of many kinds, in which a new kind of presence in the world manifested itself not merely through a series of profiles that appeared, but equally importantly through a series of profiles that might have but did not appear; its profiles were just so and not otherwise to reveal the new presence in the world. An anticipated variety of profiles were displayed in the appropriate ways in a variety of circumstances. What was manifested thus had more the character of a bodily presence than a law. Even when the discovery has been prepared for by a quite specific and rational set of procedures, the final achievement takes the form of a perceptual act of recognition.

Third, even though a recognition may come as a surprise, and be provoked by signs that are unexpected and in themselves of no consequence, recognition for Aristotle is *prepared for by a concernful, social engagement with the world*, often by a specific inquiry; recognition is not the outcome of self-inspection but worldly interaction. Oedipus is led to his fateful recognition through his relentless questioning of knowledgeable parties, while Electra's desire for vengeance over her father's death makes possible her presence at his tomb, the site of her recognition of Orestes. Hegel in particular stresses that recognition is the product of a social interaction with the world: social, because it is not the act of a solitary consciousness but is carried out within a community of other individuals; and an interaction, because the process transforms self and world. In the famous master-slave section of Hegel's *Phenomenology*, the recognition involved is even portrayed specifically as a struggle in which one risks one's life. Recognition of scientific objects is prepared for by an often demanding engagement with the world, in the form of the cultural attitude mentioned in chapter 3 to take phenomena as exemplars. I have already mentioned Dewey's point that particularly obdurate problems *(aporia)* attract more attention than others, inspire more individuals to become concerned with them, and provoke the scrutiny and reevaluation of more aspects which otherwise might have been overlooked. This engagement results in the creation of a set of experiences out of which the recognition crystallizes, so that the recognition, even of a novel structure, takes place as of something with which one is already familiar.

This, once again, is an interpretive process, which we can understand thanks to Heidegger's conception of the hermeneutical circle. Puttering around with something unfamiliar is already a moving within the circle—an interpretation, a making explicit of what I understand, a developing, deepening, and enriching of one's involvements and expectations—so that the eventual moment of confidence occurs in the form of the recognition of the presence of something with which one is *already* familiar. By the time of Röntgen's moment of confidence marking his recognition of X rays, he was already familiar with the phenomenon, even though it was a "new" discovery.

Fourth, recognition for Aristotle often involves *unanticipated or undesired consequences.* In the context of a discussion in which recognition is treated as one of the three essential parts of a dramatic plot, the unanticipated consequences naturally received considerable attention; Aristotle says that recognition is best when coinciding with a plot reversal. But numerous recognitions crop up in tragedy that do not result in reversals and are even joyous occasions, as the nurse's recognition of Odysseus through the scar on his leg. Hegel stresses especially the dynamic aspect of recognition, which for him results in a transformation of oneself as well as the world. The American sociologist Robert Merton once devoted an essay to analyzing what he called "the unanticipated consequences of purposive social action," or the ability of human action to bring about results that are unintended by the actors; these results can affect either the actors or society as a whole, and can be beneficial or not. Although Merton was specifically discussing social action, his point is equally true of scientific research.[15] That unanticipated consequences are often the outcome of scientific recognitions scarcely needs repeating; these consequences run the spectrum from the notoriously good or evil to discoveries that do not make headlines but which quietly become part of the scientific toolbox and only eventually have an impact on the densely woven fabric of scientific knowledge. Investigators frequently discover more than they wanted to or expected or could have predicted. And what is transformed by such recognition is not merely the number of phenomena in the scientific world, but the landscape of scientific knowledge itself—the "who" of the scientific community.

Too often discussions of recognition take as emblematic recognition of and by individuals, whose minds have been already equipped with whatever is needed to know the object to be recognized; the object, in turn, is considered to be preexistent and already furnished with its own identity. Aristotle's discussion, assuming a world of fixed identities, is an example. But Hegel's dynamical conception of recognition, involving the lack not only of a unique self-identity but also of the identity of the recognized object, is applicable to a perceptual account of scientific entities. For individuals in a suitably prepared community *acquire the capacity* to recognize, and the recognized object enters the world upon the achievement of this recognition. Recognition is thus stage-dependent; it is relative to the stage of development of the community whose individuals carry out acts of recognition. Recognition is a disclosure, but a disclosure to a community prepared to see.

This feature of recognition allows us to add a number of other features of recognition relevant to our inquiry: that recognition concerns the worldly presence or existence of something, is dependent on the background context, is a temporal process, is world-transforming, and involves the possibility of misrecognition.

First, recognition concerns the *existence* of something, its presence in the world. In recognizing, one's perception is occupied by the presence of the object in the world; one lingers over or is arrested by that presence, one apprehends a perceptual, worldly phenomenon *as* a phenomenon. I recognize a thing not by concordance of aspects, Merleau-Ponty says, but by the sense of completeness, by an assurance afforded to me that I grasp the order of its profiles. I know that I could be presented with more profiles than are available to me now, but I also know that I have a general idea of which. I have the sense that I will not be greatly surprised by most of them, though I could be by some.[16] By the time of Röntgen's moment of assurance, he had a general idea of how X rays would behave in certain kinds of things. Dewey claims that "in recognition there is the beginning of an act of perception, but this beginning is not allowed to serve the development of a full perception of the thing recognized. It is arrested at the point where it will serve some *other* purpose, as we recognize a man on the street in order to greet or to avoid him, not so as to see him for the sake of seeing what is there."[17] But this last is precisely the kind of recognition that I would argue takes place in basic research. In chapter 2, we encountered some reasons why Dewey would not admit this: his instrumentalist leanings, his pragmatic view of inquiry, and his lack of acceptance of scientific entities as having a bodily presence in the world.

Second, recognition is *dependent on the background context,* that is, on the state of the readable technologies through which the recognition is executed. Just as one cannot recognize an individual at a distance of half a mile without binoculars, one cannot recognize scientific phenomena without appropriate technology. Just as the binoculars make you seem close, the technology puts you at the right distance to perceive the phenomenon, so to speak. The need to be at the right distance in order to recognize something is illustrated analogously by "block portraits," which are composed entirely out of squares or other regular geometrical shape; the portraits (Abraham Lincoln and the Mona Lisa are the subjects of often reprinted ones) can be recognized only if one is positioned at the right distances from them. The spatial nearness needed to recognize the portrait is analogous to the nearness that readable technologies of the appropriate sensitivity can bring. The dependency of recognition on a background context of readable technologies whose sensitivity is often increasing provides an explanation for simultaneous discoveries, or the fact that in numerous cases inventions or discoveries were made by different individuals at virtually the same time; the background context of the community has reached a level at which a recognition, previously impossible, now becomes possible. Imagine a group of individuals watching a person emerge from a dense fog; while at one moment none of them may be able to recognize the person, the recognition might become possible to all simultaneously when the person approaches just a few feet closer. The serendipity of scientific discoveries, sometimes the subject of mystical speculations, is readily comprehensible given a sophisticated enough account of recognition.[18]

Third, recognition is a *temporal* process; it takes time to achieve. Casey speaks of *dim* and *dawning* recognition.[19] Dim recognition is never a complete recognition; it is the vague feeling that one knows someone or something but can retrieve neither name nor circumstance; Goodspeed must have felt something of this kind when he first pondered his inexplicably exposed photographic plates. It was incomplete because Goodspeed knew some phenomenon was involved, but until Röntgen's announcement he knew not which. It was nevertheless a recognition (form of knowledge), or he would not have been able to recognize in Röntgen's announcement an affinity with his own circumstance. Dawning recognition, by contrast, is the movement from dim recognition to the explicit achievement of a recognition. We must imagine that what went on in Röntgen's mind in the weeks after November 8 amounted to a dawning recognition of a new phenomenon, as he acquired more familiarity with it. Experimental checks and rechecks follow upon recognitions rather than create them. Someone will object that Arthur Koestler, for instance, discusses instances of recognition where the passage from ignorance to knowledge is supposedly *immediate*, a leap of understanding he calls the "eureka process," after the famous story of Archimedes, the putatively gold crown, and the bathtub.[20] But a glance at the story shows that the recognition was not immediate. Archimedes had spent weeks pondering the problem, time in which he became intimately familiar with its every aspect. That period of pondering (inquiry) gave rise to the set of experiences out of which the recognition crystallized. Archimedes may have cried, "Eureka!", but to call that an instantaneous acquisition of insight is naïve; Archimedes had undergone a temporal process, even though the moment of arrival may seem to have occurred instantaneously. Consider, too, Ernest Rutherford's oft-quoted remark, made late in his life, to the effect that the backwards scattering of alpha particles off of atoms, by which he recognized the presence of an atomic nucleus, was "almost as incredible as if you fired a 15-inch shell at a piece of tissue paper and it came back and hit you."[21] Here is myth-making in action; the remark would lead us to believe that Rutherford switched on the equipment, saw an unambiguous result, and immediately realized its significance. In fact, Rutherford came to realize only slowly how astounding the scattering results were, after a long experimental research program.[22] One could say that, strictly speaking, there is no such thing as instantaneous recognition.

Fourth, recognition is *world-transforming*. Recognition transforms the situation not merely in that one achieves knowledge of oneself and others or that one experiences unanticipated consequences, but also in that one has arrived at a new stage in the dialectic, characterized by a freeing from things, a liberation. Casey also discusses the transformative features of recognition.[23] In recognition, he writes, the past is supposed, the future portended. Recognition is a present-oriented act of perception, dominating the past and future experience of an object by contributing what Casey calls *availability* and *consolidation*. Availability is the name for the way in which objects become more accessible to us postrecognition; consolidation refers to the way in which the object gains stability and identity. Acts of recognition draw together numerous behaviors—intervening with screens that glow in the presence of radiation, exposure of photographic images, etc.—as belonging to the noetic aspects that exhibit different noematic profiles of the same object. At the same time, these behaviors are made available as new possible praxes. As a result, Casey says, recognition both plunges us more deeply into the world and into ourselves, into our own self-understanding. Recognition is world-transforming, the expansion of our storehouse of perceived structures of the world through inquiry in a way that cannot be undone. This is true even in cases of ordinary percep-tion, such as are involved in the recognition of block portraits. Once recog-nized, it is impossible not to perceive the new object, "as if some kind of perceptual hysteresis prevented the image from once again dissolving into an abstract pattern of squares."[24] One could say that our place in the world is defined through the kinds of things we recognize. Through education we come to recognize features of the world others already recognize. Through basic research and scholarship in the sciences and humanities novel features enter the world; new phenomena are disclosed and recog-nized (both serving to name the same thing, with *disclosed* placing the stress on the object, *recognized* on the subject).

Finally, recognition involves the *ever-present possibility of misrecognition.*

Misrecognition of scientific objects can no more be prevented than the misrecognition of individuals. It is obsessive to try to exclude it at all costs—as it would be pathologically obsessive of me, say, in ordinary circumstances to consider and exclude the possibility that an actor is impersonating my friend Fred before accepting his presence in the room. Perception always and naturally includes two circumstances that give rise to the possibility for misrecognition. On the one hand, I see only a finite set of possible profiles of what is presented, leading to the possibility of misrecognizing the structure of something (and hence that what I apprehend is an epiphenomenon). On the other hand, there is also always the possibility that a phenomenon will show itself in radically new and unexpected ways. In the former case, for instance, we could mistake electrons for muons if we see only profiles in which they appear similarly; in the latter case, electrons could turn out to be an entirely different kind of phenomenon than we thought, and what we thought were electrons could reveal themselves to be in turn profiles of another kind of phenomenon (leptons). The possibility of misrecognition is increased due to the intense competitive atmosphere of international science, one of the most competitive of human activities. Imagine different groups of individuals, in contact with each other through different degrees of personal communication or through journals, each striving to be the first to recognize new phenomena, with the ability to acquire new funds for further projects at stake. Small wonder that a few jump the gun. Misrecognition can occur in different ways, and must be distinguished from fraud.[25] Misrecognitions testify to the fallibility and perils of scientific perception, and deserve study in the philosophy of science because they are disclosive about the nature of such perception.

Recognition and the Manipulability of Profiles

The primacy of performance means here that the attainment of recognition is the transforming moment, and that while events and episodes may prepare for it, these cannot be considered the condition of the recognition. Other signs might have brought about a given act of recognition. Yet reliance on apparently arbitrary signs does not make the culminating act any less valid a recognition.

> *Countless stories from the history of science could be cited here, of spores floating accidentally through windows, inexplicably glowing screens, and so forth. A prize example involves the discovery of a new species of lemur in the Madagascar rain forest a few years ago. The discoverers, one from Germany and another from Madagascar, had used a stray dog that happened to wander into the camp as the vehicle for the discovery; the dog was held*

up to sniff holes in trees until it barked, whereupon the lemur was discov-
ered. "I don't think that is a real scientific way to bring things out," one
of the biologists told reporters.[26]

Yet the recognition was nonetheless achieved; it involved a passage from
ignorance to knowledge, a perceptual act (of ordinary perception), inquir-
ing engagement, and unanticipated consequences (puzzles about the animal's behavior that will have to be solved, etc.) And no one disputes the
scientific status of the eventual paper about the discovery, "Rediscovery of
Allocebus trichotis Gunther 1875 (Primates) in Northeast Madagascar."[27] But
the legacy of the mythic account has been to obscure the process that leads
to recognition. Witness the expressions of horror by not only members of
the public but also certain scientists that followed the publication of *The
Double Helix*. Some reviewers of that work (some of these reviews are reprinted in the Norton Critical Edition) held that the account of scientific
activity provided in it might mislead aspiring young scientists as to what
science is all about.[28] It is as if the critics felt that exposing the role of
contingent events, people, and emotions that preceded the discovery recounted in the book amounted to an attack on science. But the discovery
process did not become any less valid because things like theft of data,
skirt-chasing, and the like played roles in it.

Does this mean that recognition is ultimately based on chance? (For we
do speak of "chance recognitions," though infrequently in science.) Is the
process of science dependent upon things as arbitrary as stray dogs who
know how to bark at the right time? Not at all. Inquiry, concernful engagement, prepared the way for the lemur's recognition. One cannot say that,
because of the dog's role, the discoverers of the lemur "stumbled across"
the lemur; a process of inquiry was involved; hearing their story, one sus-
pects that they would have found the lemur somehow even without the
dog. Nor can we say that the discovery of DNA was coincidence, despite
the anecdotes and accidental encounters that shape the action of the book;
the patent relentlessness of the inquirers in their quest forbids us to take
that idea seriously. And we must imagine that Oedipus's recognition of
himself as the slayer of Laius would have transpired eventually due to the
tenacity of his inquiry even without his specific interrogations of Tiresias,
the Corinthian messenger, and Jocasta.

Recognition, then, is not a simple acknowledgement; it is an achievement, a seeing-as transformed into a new perception that opens up new
possibilities. It supplies its own standard for correctness inasmuch as it
implies an indefinite number of further profiles belonging to the object
recognized, which may themselves be sampled. Psychologists may investigate the conditions for recognition; historians may track down the chronol-
ogy leading up to moments of recognition; sociologists may point out social
elements that facilitated them. But, philosophically speaking, recognition

can be identified as a phenomenon in its own right, possessing a certain kind of structure. No rules can be formulated for infallibly recognizing novel phenomena, for recognition depends on an ever-changing background context and the relative novelty of what comes to be recognized. (What happens in the case of optical scanners and other instruments involving pattern recognition is that the background and types of foreground figures have been standardized.) When the background is not standardized and one wants advice on how to achieve a recognition it is difficult to improve upon the advice of Freud, one of the great attenders to performances, to pay "evenly hovering attention."[29]

Achieving a recognition does not depend on the manipulability of the object recognized. This is relevant to the question of whether or not there is a priority of some of the sciences; whether, for instance, physics, chemistry, and biology are more "scientific" than, say, geology or ecology. It may seem that the former group have a superiority to the latter group because their objects are manipulable to a greater degree. Again, however, one must recall the primacy of the phenomenon. The cultural attitude of science, I said in chapter 3, is to seek to take objects and events as profiles of phenomena. And recognition, the name for the process in which something is apprehended as a profile of a phenomenon, may take place whether we can manipulate the phenomenon so that all of its profiles appear at will in a laboratory, or have only limited access to profiles in the field. I can recognize something as a desk by walking around it or by seeing it at a distance; I can know something is a mountain even if I have never climbed to the top or have only one limited view of it. This makes recognitions that are achieved by laboratory manipulations (as in physics, chemistry, and biology, say) are on an equal par scientifically with those achieved through observations made in the field (as in geology, paleontology, oceanography, and ecology). The serendipitous character of discoveries of fossils of dinosaurs hatching from eggs, of teething dinosaur embryos, of dinosaurs who had just eaten something before being interred in tar pits, does not make the phenomena recognized in these discoveries any less scientific.

Ernst Mayr has decried the so-called methodological distinction often drawn between experiment on the one hand, and comparison and observation on the other, and the utilization of such a distinction to mark an important difference in status between physical and biological research. Mayr ardently defends comparison and observation, and champions the validity of the conclusions that they yield.[30] But given the primacy of performance, this distinction becomes unimportant and Mayr's defensiveness unnecessary. What matters is recognition, and whether it transpires through the encounter with profiles produced in the laboratory, or through profiles that appear in the field is unimportant.

Scientific phenomena may also have profiles that appear in nonscientific contexts. Supernovas and other astronomical events have been recognized in certain apparitions depicted in ancient pottery, which helps archaeologists in establishing time lines. Thixatropy, or the phenomenon that certain viscous gels liquify upon being stirred or shaken, may well be a profile of the same phenomenon as that responsible for the liquification of the clotted blood of saints on ceremonial occasions, as the blood of Saint Januarius in Naples does every few months. And scientists have described certain rare meteorological phenomena bearing an uncanny resemblance to the descriptions of certain alleged visions in scripture.[31] No proof is possible that such visions were indeed ordinary perceptions of intramundane meteorological phenomena, of course, but the possibility remains of accounting for such visions in that way. In cases such as this, the cultural attitude of science, to take objects and events as profiles of phenomena, clashes with other cultural attitudes.

VII

PERFORMANCE AND PRODUCTION
THE RELATION BETWEEN
SCIENCE AS INQUIRY AND
SCIENCE AS CULTURAL PRACTICE

At the end of chapter 3, I said that the absence of experimentation from our picture of science could be filled in with the help of the theatrical analogy, or the point-by-point comparison between scientific experimentation and theatrical performance. Performance has three principal dimensions: presentation, representation, and recognition, and analogues of each of these were discovered in experimentation and elaborated in chapters 4 through 6 with tools provided by the three thinkers discussed in chapter 2.

The theatrical analogy has allowed us to redefine experimentation philosophically. Experimentation is a process of inquiry that seeks to make phenomena known through the performance of actions. Performances are executed by and for members of a *suitably prepared community*, in response to *problematic situations*. *Reconstruction* of those problematic situations gives rise to a more assured, deepened, and enriched engagement with the world. Inquiry is *interpretive*, involving the development of the understanding via moving in the *hermeneutical circle*, two versions of which may be distinguished: *text hermeneutics*, involving textual interpretation, and *act hermeneutics*, involving the performance of actions. Problematic situations are reconstructed via recognition of phenomena that appear in performance; taking objects and events as instances of phenomena is the *cultural attitude* of science.

Performances have three principal and related dimensions: (1) they are skillfully executed actions, or *presentations;* (2) they are actions in which a suitably prepared community seeks the *recognition* of phenomena; and (3) both the performance process and the performance product are structured by a *representation*.

When successful, experimentation involves the appearance of *phenomena;* experiments *prepare* phenomena for study. Phenomena are accessible to *perception;* scientific entities are usually (but not always) accessible to perception via *readable technologies*. Phenomena appear through *profiles*, which particular profile depending on the relative positioning of observer and observed. The regularity of profiles under passive and active transformations is the *invariance* of the phenomenon. An invariant names an identity of structures in subject and object; it is the noetic-noematic correlation. Invariants entail *horizons* of possible profiles, which are given together with the phenomenon; in exploring phenomena in inquiry, horizons are constituted, filled in, revised, and extended. Horizons may be *internal* or *external*. One acquires confidence that one has accurately represented a phenomenon through an ability to move from profile to profile. Expectations may be *fulfilled* or not, for a phenomenon is capable of revealing itself in new and unanticipated ways, and the invariant structure of the perceived phenomenon may have to be adjusted accordingly. A scientific phenomenon is something "behind" the data that *denominate* its presence and "behind" the theories by which it is represented. Representation is related to the kind of thing a phenomenon is, presentation to an individual profile. Phenomena can be *classical* or *path-dependent*.

The *principle of the primacy of the performance* means that presentation, representation, and recognition are each in the service of the appearing of the phenomenon; data are *fluid*, theories *fragile*, and recognitions *stage-dependent*. For not all interpretations of the world are contextually legitimated; a performance can exceed the theory used to create and understand it, calling for new performances and theories. The unity of the sciences, and relations between different branches, are a function of the unity of and interrelations between the phenomena involved.

A host of objections may be raised. Someone will object that scientific performances concern a world already made, theatrical performances a self-created one. But the difference is due to different kinds of phenomena presented in performance, not to the basic performance structure itself. Someone will object that theatrical performances are more "subjective" than experimental ones due to the presence of theory and thus an objectivity in the latter. But the representation in each case is a script whose function is to allow us to come face to face with the things themselves in performance.

Someone will produce a sample of book reviews or opening night reviews of famous musical or theatre works to try to score the point that theatre criticism has proven remarkably inconsistent over the years even in the hands of renowned commentators, and object that in science much more objective criteria for the evaluation of experiments is available. But when the background context is standardized with respect to equipment, techniques, expectations, ways of recording and so forth, there are indeed shared procedures within the scientific community for evaluating what is a good experiment. But these are not always present, and in that event a play of perspectives will produce controversy as each individual relies on a different experience, acquaintance with procedures, advice of friends,

hunches about proper methods and direction of the field, and so on. Anyone who doubts this should attend a meeting of a laboratory program
committee faced with the task of selecting from among several proposals
the minority to be supported by the lab. Fierce competition often emerges
among noted scientists and vast disagreement over the value of each proposal. Shouting matches have transpired between eminent researchers over
the value of new and unstandardized techniques such as magnetoencepha-
lography.[1] Similarly, while opening night reviews of a new work are one
thing, and subject to the idiosyncracies of the individual reviewers, there
are cases where a standardized background context exists—say, at a graduate school evaluation or recital. In that case, it is possible to put aside
one's temperament or personal inclinations and achieve a consensus.

Experimental performances thus lead to the disclosure of new entities
in laboratory and other situations. These new entities may be related to
already familiar entities; the laboratory phenomenon of electrons may be
related to lightning, for instance, or the laboratory phenomenon of optical
refraction and reflection may be related to rainbows. But these familiar
entities are not deductions from the theory in an explanatory sense. When
one understands the scientific account of a rainbow, one experiences something new—an account of the behavior of light under certain conditions—
that is now coupled with the familiar experience of a rainbow as a multicolored thing in a hermeneutical way. When I understand the scientific account of the rainbow, I do not now see the scientific rainbow instead of
the life-world rainbow, but merge the two—I see the multicolored rainbow
as an instance of something else; I see the multicolored object as a presentation or profile of another kind of phenomenon, namely, of the reflection
and refraction of light rays in droplets of water.

Given a sufficient level of abstraction, then, both theatrical and scientific
activity exhibit a similar structure involving an interaction between world,
performance, and audience. Just as the dramatic world changes in part by
critical evaluation of performances and in part by changes in the external
horizon, eliciting a demand for new scripts and new theatrical performances, so the scientific world changes in part by evaluation of experimental
performances and also in part by changes in the external horizon, eliciting
a demand for new theories and new experimental performances. Vitality
is defined not by any final achievements in this process of interaction, but
by continued motion.

Production

While a general characterization of performance can be attained by addressing its three principal dimensions, another important structure is
addressed by *production*.[2] A phenomenon does not lend itself to all circum-

stances, and lends itself differently to different circumstances; it *belongs* to different circumstances differently. Röntgen's X rays present themselves through glowing screens in some circumstances and exposed photographic plates in others; the rays are, we might say, differently *produced*. If one somehow were able to "replay the tape" of this episode, thoroughly erase everything to a point in time a number of years prior to November of 1895, and let the tape run again with all the decisions and actions free to occur differently, X rays might well have presented themselves to someone other than Röntgen in a different form. Production thus plays a crucial role in experimental inquiry.

In theatre, the need for a production arises from the fact that, by itself, a script or score is but an abstract testimonial to the possibility of a performance. To realize a performance, a number of decisions have to be made and acts accomplished in advance to specify and create the circumstances. Advance decisions involve aspects like casting, props, staging, blocking, and costumes; advance actions include securing funding, obtaining a theatre, and drawing up a production schedule.

Similarly, theories are but abstract testimonials to the possibility of the presence of phenomena. Theories do not specify, for example, when or where a phenomenon is to take place in a lab. The process of experimental production forces a researcher to make a number of decisions and to accomplish a number of actions in advance of an actual experiment in order to create the environment or special context in which the phenomenon may appear. Advance decisions involve choice of collaborators, site, equipment, timetable, and design; advance actions include securing funding, necessary permissions, lining up contractors if necessary, and drawing up organizational plans and schedules. This environment or special context then provides a stable content to the runs or performances of that experiment, making it possible to speak of repeating runs and even of repeating experiments; the latter would occur if, elsewhere, the same set of decisions regarding the phenomenon were to be made and the same environment recreated. (In theatre, too, specific productions of a show as opposed to the show itself can be revived in a similar manner.)

And just as there may be different "runs" of one preparation, there may be different ways of "producing" the same phenomenon. I mentioned the various ways Röntgen produced his X rays; in modern physics the productions can be much larger in scale. The Glashow-Weinberg-Salem electroweak theory, for instance, lent itself to several different kinds of productions: of neutral currents, of the scattering of polarized electrons, and of atomic parity violation effects. These amounted to three different ways of producing the same phenomenon (through any number of individual performances), the electroweak force, on the basis of a single theory.

Production requires one to operate in a concrete social context, and any event "produced" therefore is shaped by political, environmental, economic, and psychological factors. Even deciding to build a laboratory or observatory in a specific place is the outcome of such factors. The site selection for a modern scientific laboratory is inevitably a protracted and highly political process. A long list of scientific, environmental, and political factors were taken into account in the planning of Brookhaven National Laboratory, for instance, incuding the consideration that the new laboratory, the first in the Northeast region of the United States with a nuclear reactor, be an overnight train ride from the universities involved in its planning.[3] Once constructed and commissioned, laboratories then become part of the "world" of science, part of the available sites where certain kinds of experimental performances can take place. At this point, other sets of social factors come into play in the practice of science, each of which could become the theme of an inquiry in its own right. At any given time, thousands of experiments might be performed; national laboratories and observatories are swamped with proposals for experiments, and can accommodate only a fraction of the requests for lab time put to them. The choice is dictated by a variety of motives, including the reputations of the scientists who make the proposals (including how qualified, reliable, and easy to work with they are), the facilities of the laboratories, the aims of the laboratories, the ability of a team to draw outside funding, and other factors such as the desirability of international collaboration.

Not all sites may be available or have the appropriate facilities. Anyone wishing to arrange a series of astronomical observations, for instance, has a choice of observatories to approach with different capabilities; moreover, these observatories are generally oversubscribed and one may be forced to work with a less convenient site to gain time on a telescope. Similarly, someone wishing to produce an experiment in high-energy physics has only a small number of laboratories with large particle accelerators from which to choose, and even if one's proposal is approved one must compete for beam time with other ongoing experiments. Here again, a less convenient site may be more available. One factor, however, is inevitably a judgment about the pressing scientific questions of the day and which of the proposed experiments best addresses them.

An experimental production also embodies an entire set of judgments about how that phenomenon is to appear in an experimental performance. The Glashow-Weinberg-Salem electroweak theory was a "script" with a variety of possible performances; experimental productions could involve vastly different contexts in which these performances might take place (national laboratories, or at the bottom of deep mines), each of which called for vastly different instrumentation, funding, and personnel. The simplest Grand Unified Theory, on the other hand, was a script with few possible performances given present-day instrumentation—one performance being proton decay. Several groups of scientists working in different productions proved unable to execute performances in these environments, suggesting that the script had to be revised.

Moreover, in producing an experiment one must choose and work with a set of available or promising technologies, a set of personalities, available money, contracts and regulations, and so forth. Questions arise: Can it be done cheaply enough? Can these people work well together? Is the schedule realistic? Are supplies available? Are the techniques reliable enough? Can one work within the prevailing regulations and restrictions—involving, for instance, handling and disposal of radioactive materials, union regulations, research on human or animal subjects?

Production may require considerable ingenuity. It is desirable when using lead shielding for cosmic ray experiments, for instance, to have the lead refined to eliminate naturally occurring radioactive isotopes. Refining is expensive, however, and clever experimenters seeking to bring in their projects under budget have from time to time turned to other sources, including old lead coffins, and lead ingots from the cargo of ancient Roman sailing vessels. This illustrates ingenuity in production for budgetary reasons; ingenuity is also frequently needed to create the very conditions in which a phenomenon will appear. Such ingenuity characterizes a good experimenter, making it possible for that experimenter to "reach" a result before someone else. Consider, for instance, the cosmic ray studies based on analysis of rat urine. The middens or nests of desert rats consist of a jumble of sticks, garbage, and other material that then becomes drenched in the animals' urine. Such middens have been preserved for thousands and even tens of thousands of years, and are treasure troves for biologists studying plant life thousands of years ago. They have also been used for cosmic ray studies. Cosmic rays passing through the atmosphere change argon atoms to chlorine-36, which is then taken up by plants, consumed by animals, and discharged in urine; the amount of chlorine-36 in urine is thus an indicator of the cosmic ray flux. By examining the chlorine-36 content at various levels of a rat midden, scientists have been able to obtain a reading on the cosmic ray flux at various points in the Earth's recent history. In this case, it is not a matter of a clever production decision (in creating the right environment in which to seek profiles of a phenomenon) making the appearance of a phenomenon to scientists economical—it is a matter of a clever production decision devising the conditions in which the phenomenon can appear at all.

Personal style is often involved in production; it was a factor, for instance, in the choice of which large experiments to construct at the Superconducting Supercollider.[4] Production is also the domain examined by the sociology of science, such as whether scientific activity is normed by particular kinds of values, presuppositions, and mores. One may get the impression nowadays that things like competition, personalities, patents, and publicity exist only in bad science, but they have a formative role in good science as well. It is important, however, to see their role in the experimental inquiry; namely, in production, or in the set of decisions and actions that have to be executed before experimental performances take place.

Do anecdotes about the role of such things as ingenuity, personality, and competition truly reveal something essential about science? Only someone hopelessly naïve about scientific activity can answer in the negative. Researchers are not interchangeable parts so that it matters little whether A or B conducts an experiment. Significant differences can occur in experimental performances because A did it in his or her way, and B did it in another way. The final performance or "product" is dependent on the process that led to it; yet neither does it reduce to the cultural and historical influences on that process. Therefore, unless one knows how to address the role of something like ingenuity or personal style in production, one is not yet in possession of a serious or adequate account of experimental inquiry, and hence of a true philosophy of science.

Production is one of the most prominent aspects of the large and complex contemporary high energy physics experiments, and the requirements of such a production place constraints on the kind of performance that ultimately takes place. Furthermore, experiments are influenced by what might be called production values, or factors to be emphasized in performance. In the experiments to determine whether cold fusion was really a phenomenon, for instance, researchers in the community came to direct their attention to the role of certain key ingredients in descriptions of the experimental performances to judge whether the profile was really present, including measurements of the amount of tritium, neutron emissions, and temperature rise. Experimenters thus took special steps to convince the community of the reliability of the instruments and methods used to make such measurements.

The role of production allows us to resolve the philosophical problem of what I called the antinomic character of scientific knowledge: the fact that, on the one hand, it is affected by the historical, cultural, and social contexts in which it develops, but that, on the other hand, it also has a kind of objectivity or independence from these contexts. Social constructivists stress the dependence of science on historical, cultural, and social contexts, while positivists and realists stress its objectivity. Both positions, I said, have validity; human beings "make" science but not any way they please. The task of the philosophy of science is to model scientific activity in a way which illuminates this dual character; otherwise, the argument between partisans of each position will continue indefinitely.

The theatrical analogy allows us to create just such a model, in the difference between performance and production. While experimental *productions* always take place within a social and historical context, within an external horizon, and thus bear the imprint of the various forces at work in that context (imprints that can be empirically studied), experimental *performances* reveal phenomena having a measure of independence from that context insofar as they reveal themselves as having profiles in other kinds of contexts. The concepts of performance and production thus allow a better understanding of the interrelation of science as inquiry and science as cultural practice.

Thus, to our list of definitions at the beginning of the chapter, we must add that performances are *produced* by a set of decisions and actions executed in advance of the performance. The *antinomic* character of performances of any sort is that they are simultaneously *ontological,* or concerned with the real presence and disclosure of invariants in the world, and *praxical,* or shaped by human cultural and historical forces. The antinomic character of science gives rise to the temptation to overemphasize one of two different aspects; its objectivity (invariant structure) on the one hand, and its social construction on the other. Both temptations should be resisted, for performance involves the coworking of each.

Any appropriately structured inquiry into science at some point must identify the various social, historical, cultural, and economic factors involved in the practice of science, and locate their proper place. This inquiry uses the theatrical analogy as a preliminary way of recognizing these factors, and to model their interaction. It would be as much of an error to *ignore* the presence of such factors as it would be to *reduce* the activity of science to them. The traditional philosophy of science has taken the former path, supposing that social, cultural, and historical influences either do not exist or that they are unimportant. Social constructivists and epistemological relativists, on the other hand, have chosen the latter path, considering socio-historical-cultural factors to be determinative. They underplay the constraints placed on the development of scientific experimentation and inquiry by the necessity to achieve the real presence of scientific phenomena in the laboratory as achieved through readable technologies. The theatrical analogy exposes the naïveté of both approaches. The unsophistication of someone who asserts that the theatrical world is exclusively about explorations of the Meaning of Human Existence is as evident as the cynicism of someone who claims that it is all politics or personalities. The point, of course, is that both views are right to an extent. The theatrical analogy, in conjunction with the schema of pragmatic hermeneutical phenomenology, allows us to *perceive* the situation aright, to *reconstitute* our perception of scientific experimentation, and to notice that scientific phenomena, like those of theatre, takes place amid a complex interaction of internal and external horizons. Scientific experimentation is at once ontological and praxical, indebted alike to Being on the one hand and history and culture on the other.

The ideas pertaining to production and performance outlined above allow us to describe conveniently the existence of different ways in which a scientific research program may be conducted. First of all, it may be

directed to looking for the theory behind a series of repeatable performances—performances that experimenters know how to prepare consistently. Röntgen's efforts (and those of his contemporaries) to understand
what was making his scintillation screen glow is an example of this kind
of research program. In such research, theorists are looking to write the
theory or script, as it were, for a series of performances that have been
consistently executed. A successful writing of the script presumably would
show the possibility of performances in other circumstances, and possibly
lead to their standardization.

Second, scientific research may be directed to preparing a performance
for the first time. In more traditional language, this would be "discovering"
a phenomenon predicted by theory. Seeking a subatomic particle predicted
by theory is an example of this kind of research. Here, experimenters are
looking to see whether a certain performance can be achieved, so to speak,
on the basis of an existing script.

Third, scientific research may be directed to seeking out deviations between experimental performances and theoretical scripts. Experimenters
or theorists alike may engage in this kind of research program. Experimenters may try to prepare performances in novel conditions to see
whether such performances still "fit the script." Instances of this include
measurements of familiar phenomena, such as the position of hydrogen
spectral lines or important constants, to an unprecedented degree of accu-
racy, or the half-century-long quest to discover expected breakdowns in
the theory of quantum electrodynamics at close range. Theorists may try to
see in already executed experimental performances indications of another
"script" and attempt to write it; the Eightfold Way is an example. Here,
the theatrical analogy is in parlous shape; it would be as if directors continually tested the limits of how a play can be performed while critics
sought to pick out novel features of the performance not codified in the
script—and if, furthermore, this information required scripts to be reworked.

Science as Inquiry and as Cultural Practice

The model of scientific activity as involving the distinction between performance and production also helps to signal the dangers of overstressing
aspects of experimentation. One mistake in an inquiry into theatre, for
instance, is to focus too much on the script, viewing the performance as
a species of demonstration or ornamentation. Equally misguided is the
approach that focuses too much on the cultural and historical context in
explaining the meaning of a performance. Imagine a drama critic able to
account for each and every detail of performance of, say, *Julius Caesar*, as
a function of political and social factors. Although the army uniforms and

boots worn by the Roman actors in Orson Welles's 1935 production can be explained as part of the theatre world's horrified reaction to the spectre of fascism then rising over Europe, it would be a mistake to reduce the production to an antifascist diatribe. Some critics did so, of course. But Welles's *Julius Caesar* was a *play*, and not only a sort of theatrical op-ed piece responding to the specific political context; indeed, it was the *same* play that had appeared in many other political contexts and been spoken in a different way in each.

Similarly, science is not just theory, for theories are fragile; they are theories *of* performances and the aim is to represent the phenomenon that puts in an appearance in performance. Nor is science only data, for data are fluid; data are descriptions of the way something appears in a particular context, and science is interested in the phenomenon that appears rather than merely how it appeared. Science is not just the production of literary texts, as "laboratory studies" people often would have it, for these texts are accounts of performances or the preparation for such performances. Science is not just about the domination and control of nature, for performances involve not mastery or control but play. Nature is not infinitely pliable, not all performances are possible, and one must engage Nature to "play along" in order to discover the rules of that play. Science is not just about economic or political praxes, as social studies of science scholars occasionally imply, nor is it about the clash of ambitious personalities, as some journalism would have it, because these relate only to the social dimension involved in the preparation of performances; experimentation is a *poiesis*, or a bringing-forth of some phenomenon through praxes. One could just as legitimately claim that theatre is about box office, or the clash of ambitious personalities, or the desire for fame or power, and so forth. Social forces have their place in the appearing of phenomena in performance, but if human beings were not interested, fascinated, and preoccupied by the performances, they would not happen. If we view scientific activity without the productive aspect, as positivism attempted to do, then we have no understanding of the role of social and historical forces at work in it. If, on the other hand, we view science without the performance aspect and concentrate wholly on the productive aspect, as the social studies of science scholars often do, then we are in danger of seeing in science only the arbitrary clash of forces. It would be like flying over a soccer game in an airplane sufficiently high up so that one can see the competitors but not the ball; the players will seem to ebb and flow in a series of interesting behaviors exhibiting many different patterns—patterns that could be described empirically in great detail—but the key element that would allow us to grasp the real meaning of the game would be invisible.

Consider, as an example, the light that the performance-production model sheds on the controversy over the nature, desirability, and dangers of "Big Science," which involves scientific production. Alvin Weinberg, then director of Oak Ridge National Laboratory, coined the phrase in a 1961 article entitled "Impact of Large-Scale Science on the United States."[5] Ever since, "Big Science" has been a standard term in the lexicon of those who write about science, though not always with the same connotations. While Weinberg, for instance, was cautionary about the prospect and stressed the dangers of large scientific projects, others were enthusiastic and emphasized the opportunities. Today, "Big Science" is generally used as a term of opprobrium.

Large scientific projects pose certain dangers, Weinberg wrote, including "moneyitis" (expending money not thought), "journalitis" (public rather than scientific debate on projects), and "administratitis" (an overabundance of administrators). He criticized the manned space program for "hazard, expense, and relevance," and was unenthusiastic about large accelerators, which were more scientifically valid but equally remote from human concerns. He wondered whether such projects would sap resources of science and society, and proposed redirecting money to "scientific issues which bear more directly on human well-being." Weinberg's aim, however, was not to cast judgment but to inaugurate "philosophic debate on the problems of scientific choice."[6] Big Science, he felt, introduces new issues into science policy that must be exposed and addressed lest they be settled by default at the expense of scientific "productivity" (note that this word has a different meaning from what I have called "production," but the use of both terms in this chapter is unavoidable). But additional issues have appeared in the intervening years, and the debate over the value of Big Science and its impact on productivity has continued unabated.[7]

Productivity is notoriously difficult to measure and even define, regardless of the field. Consider agriculture, for instance, where the definition and determination of productivity might seem, erroneously, to be relatively straightforward.[8] To take a simplistic example, a farmer faced with a choice of what to plant on a particular plot of land could decide to maximize monetary profit, number of calories per acre, amount of protein per acre, security of the harvest, number of calories per man-hour of labor, prestige of the farm, and so forth. There is, in short, no single index of productivity. Each option mentioned is guided by a different set of possible values which puts into play a different index of productivity and suggests a different crop. In practice, of course, no sole value would likely be given entire priority and the actual outcome would be some compromise.

In science, the matter is further complicated because, Weinberg says, the "product," the understanding of and ability to manipulate nature, can be evaluated by two different kinds of measures which he called "internal" and "external" criteria. Internal criteria "arise from within the science itself, or from its social structure and organization," while external criteria "stem from the social or other setting in which the science is embedded."[9]

Neither kind, in turn, involves a single index of productivity; within each set different possible values can be identified implying different indices.

In practice, as the working scientist knows only too well, the decision of which scientific projects to support is the outcome of a highly political process generally involving compromises between a number of different internal and external values. Moreover, the social negotiation involved takes place on a number of different levels. Science, for instance, competes with a number of other activities that also are perceived to be of some economic, military, cultural, or political value. Within science, in turn, a competition exists for support among its different branches, such as particle physics, oceanography, astronomy, chemistry, molecular biology, and geology, among others, each with its own perceived value. The competition continues within each branch, where different projects are rivals for available resources.

The Big vs. Little Science debate, which is all about that aspect of scientific activity that I have called production, was spawned by the fear that the emergence of large scientific projects threatens to skew an otherwise healthy competition on all levels and to distort the way values are applied to evaluate projects. A large project in one branch, it was felt, could soak up money that might be shared by several smaller but equally valuable projects. Moreover, a large project in one branch might get out of hand and wind up unfairly expropriating resources otherwise destined for other branches—or even for worthy nonscientific activities. In recent years, the percentage of the total research and development budget consumed by largest projects has increased.[10] It is undeniable, as Weinberg foresaw, that this development has changed the way scientific experiments are conducted, and the conventional wisdom is that it has brought about the impact of what Weinberg called external values on their planning and execution.[11]

But the matter can be elaborated in a clearer way, I think, by reformulating Weinberg's distinction as that between science as *performance* and as *production*. An experiment, I have argued, is a kind of performance, understood in the broadest sense of an action executed to see what happens in order to satisfy an interest. In science, the actions are those of instruments interacting with nature, and the interest is connected with a specific inquiry into natural structures. The performance values of science are those that promote the skilled execution of experiments, and include how well an experiment is thought out, the quality of the investigators, and the relevance of the experiment to the principal directions of the field. Production, on the other hand, refers to the interaction between planners and the particular social, political, technological, and economic context required that a performance may take place. Production values of science can include social and economic returns for society, improved instrumentation, international cooperation, and national prestige. The distinction between performance and production values in science is crucial and must be born in mind at a time when so much of science threatens to dissolve into politics.

But it is misleading to imply, as Weinberg does, that production is "external" to science, given the essential place of production in scientific activity. Moreover, more performance values exist than the two ("ripeness for exploitation" and "caliber of the practitioners") Weinberg mentions as internal, and a wider range of possible production values than the three external species he identifies (technological merit, social merit, and scientific merit).

One issue highlighted by the performance-production distinction is the existence of different models for Big Science in various areas involving dramatically different relations between production and performance. The Superconducting Supercollider, for instance, is a large instrument serving a relatively small number of experiments with low diversity; synchrotron radiation facilities provide centralized staging areas for numerous small maximally diverse experiments; the genome project is a noncentralized coordination of smaller efforts. Optical telescopes are another special case, due to the availability of private money. The community spectrum served, the kinds of risks, and the potential returns are so varied as to involve in each case a different kind of productivity—and a different meaning for "Big Science."

A second issue highlighted by this distinction involves risks that accrue from the fact that the time it takes to complete present-day productions can be so extended—over a decade—that interim changes in the scientific world can alter the productivity of the eventual experiments. The speed, quality, and relevance of a certain kind of experimental production may change in the time it takes to complete one, possibly rendering it obsolete. The factors involved may be of three sorts: technological breakthroughs, completion of other projects, and new information. In the years since construction began on the Hubble Space Telescope, for instance, developments in adaptive optics increased the resolution of ground-based observatories, other "windows" have been opened in the electromagnetic spectrum, and the general body of astronomical knowledge have all changed, forcing changes in the original estimations of the productivity of the device.

Third, the increased size of productions means increased government involvement not merely because the more resources a society has to shell out for them means a greater expectation of return, but because of a greater social interest in the way the interactions are handled. Larger productions attract more attention to the potential impact on the environment, considerations of national security and industrial competitiveness, accountability and the importance of guarding against such things as fraud, collusion, inefficiency, and so forth. Moreover, the larger the scale of a production the greater the temptation to use it as a vehicle for advancing social ends;

governmental institutions may insist, for instance, that scientific projects follow "Buy American" and minority business provisions.[12]

Fourth, the realization of a production might have social spinoffs that must be distinguished from the spinoffs of scientific knowledge itself. Technologies may have to be developed or created in the construction of a production that can be successfully transferred out of the laboratory. Constructing a state-of-the-art particle detector, for instance, is an immense production that forces detector physicists, in order to create an instrument that would be at the cutting edge for the maximum period of time, to develop new technologies. In the course of the construction of one particle detector a number of years ago, scientists taught a company that made, among other things, teddy bear whiskers how to make high-precision plastics needed for the detector in exchange for an economical rate; the productive skills acquired by that company in the process then allowed it to compete successfully for military contracts. Some attempts have been made to try to quantify spinoffs arising from the construction of high-energy physics contracts.[13] This kind of production-related spinoff is to be contrasted with performance-related spinoffs that are an outcome of the knowledge gained—for example, the discovery of the X ray, laser, and fission.

Finally, the aims of a production may not be fulfilled by the performances. It has often been the case that the technological implications of the most important and far-reaching discoveries, most notoriously those of the X ray, nuclear fission, and laser, have had nothing to do with the aims of the research programs in which they were first encountered. A similar comment could be made regarding scientific merit; while in some cases discoveries and developments in one field do find immediate use in neighboring branches, in other instances the applications come unexpectedly from far afield. The same is even true of the social value of a project; many of the breakthroughs in the "war on cancer" came not from projects targeted specifically for that purpose by President Richard Nixon's legislation, but from various and apparently unrelated work, including research on yeast, *Xenopus, Drosophila,* and *Caenorhabditis elegans.* In retrospect, it is fortunate that funds for such projects had not been diverted to the war on cancer effort. Undertaking a production—e.g., a war on cancer, on AIDS, on high-tech space defenses—does not guarantee that the ambition will be fulfilled.

Developing the concept of production may thus help to clarify many issues involved in Weinberg's "philosophic debate on the problems of scientific choice" by allowing us to recognize more features of the process of preparing and executing an experiment than emerge in most discussions of the issue. Like the general analogy between the sciences and the theatrical arts of which it is a part, the analogy with production helps guide development of a language with which to speak about experimental activity that enables one to assign a place both to the cultural and historical contexts that influence experimental activity (and which, for instance, are studied by social constructivists) and at the same time to the invariants that show through such contexts in that activity (on which postivists and scientists themselves rightly place so much emphasis). The analogy helps to show how scientific activity can both exhibit the presence of social factors without being reducible to it. The result is to clarify the much-misunderstood relation between science as inquiry and science as cultural practice. Thus, the benefit of replacing Weinberg's distinction between "internal" and "external" criteria with that between performance and production is not merely that a few nuances are added, but that the new distinction brings the problem in question within the purview of a more comprehensive picture of science itself.

Implications for Narratives about Science

Philosophers have tended to hold storytelling, or the organization of material about a subject into a single descriptive episode following roughly chronological order, in disrepute since the time of Plato. Plato's objection was that storytellers (actors) are imitators and thus one step removed from the reality they are reflecting. Moreover, in imitating one must heed appearance rather than substance and cater to one's audience, so that the product is not even an adequate imitation but a distortion rather than truth. Plato's argument still exerts force today, especially among so-called "new historians" who include Marxists, practitioners of the American cliometric methodology, and members of the French *Annales* school. These groups disown storytelling, shunning descriptions of the particular and concrete in favor of "scientific" methods allegedly able to yield more universal and eternal truths.[14] The activity of the storyteller seems in contrast to be but a pale echo of truth rather than a discovery or creation of it. The storyteller appears to be in the position of playing Aaron to Moses, passing on an already disclosed truth albeit in a form more readily comprehensible to the public. Like Moses, the subject of the tale told by the storyteller (who could be a primary lawgiver, explorer, religious figure, artist, or scientist) has one foot in the sphere of the divine, participating in primordial disclosure, bringing to ordinary mortals in the world some previously undisclosed knowledge from the beyond. The storyteller, like Aaron, seems relegated to the role of amanuensis or mouthpiece, the person who lives first of all in the mundane world and who interprets primordial activity so as to make it accessible to the public, but is able to do so only by using distortions, mediations, corruptions, descriptive metaphors, popular language.

This attitude among historians has its counterpart in a particular breed of new science history practiced by social constructivists. Just as advocates of the "scientific" methods mentioned above, which are ultimately of positivist inspiration, tend toward a determinist view of history with an emphasis on social and institutional factors, on the impersonal forces of demography, on the leading role of economics and politics, and so forth, while underplaying the role of the culture of the group and of the wills of the group and individuals, so these new approaches to science history also tend toward determinism, emphasizing the role of technology, class, social, political, and economic factors while underplaying the role of individuals, the contributions of nature, and the impact of character and chance.

Recently a renewed appreciation for the value of narrative among historians has appeared.[15] The new appreciation was prompted by the awareness that narrative is a tool able to disclose the "event-character" of human life in a way available to no other mode of presentation. As the organization of information into a roughly sequential order exhibiting the decisions affecting a path-dependent phenomenon, a narrative is able to relate the contingent set of decisions actually made in a production with the appearance of a phenomenon that appears in and through that production. A narrative is ideal for exhibiting, in other words, a path-dependent nonclassical phenomenon because it presents the evolution of its appearance along with the contingent decisions that gave rise to that evolution. The previous reflections on the nature of experimentation suggest the fruitfulness of the narrative technique for understanding it.

A fully told story of an experiment, for instance, might involve the weaving together of several different story lines.[16] These include: (1) a story of science itself, and why certain areas of science (weak interaction physics or nuclear cross sections, for instance) were seen as more crucial to pursue, more authoritative, than others; (2) a story of the instruments used in this pursuit, each of which having its own story of development and production; and (3) a story of individuals who conceived, produced, and executed the experiment, and how each of them came to learn what the important problems were and how they came to anticipate the solutions they did. These are only the principal story lines; others include the stories of the various experimental techniques involved (bubble chambers, neutron scattering, etc.) and the stories of the laboratories where the experiment is conducted. One can pursue separately one or more of these story lines, of course. But a true narrative attempts to incorporate each, for as each evolved so did the experiment.

A narrative about a discovery made with a cloud chamber—of the meson, say—might focus on technical details of the apparatus used by the three teams that discovered it almost simultaneously. Or it might focus on production-related factors such as the cultural and historical forces which led to the development of cloud chambers, the institutions whose researchers were given the freedom to pursue such studies, or the journals whose different publication demands determined the order of publication of the discovery papers. Or it might focus on the personalities and actions of the individual researchers. Each of these provides a legitimate perspective for writing a discovery account, for any discovery made with a cloud chamber is intelligible only as disclosive of nature, within complex historical space, and as the act of human beings.

But it would be a mistake to limit the possibility of an account to one of these perspectives; the "event-character" of the discovery process emerges only when each of these perspectives are preserved. It would be as if one tried to tell the story of the assassination of Francis Ferdinand *only* in terms of the detonation of a charge in Gavrilo Princip's gun, the trajectory of the bullet, and its interference with vital life processes inside the archduke; or *only* in terms of Serbian nationalism; or *only* in terms of Princip's personal motives. A first implication of the previous chapters for narratives about science is thus that while narratives can be told about science that are located in one or more particular perspectives, such as individuals, science, institutions, equipment, and production, science itself transpires through the intertwining of all of them.

But there is a deeper implication, I think, having to do not with the content towards which the attention of the science historian is drawn but with the manner of execution of the narrative itself. The construction of a narrative is itself an act carried out for the purpose of disclosing something about science, allowing it to be witnessed for its own sake. This suggests one further argumentative analogy—that narratives are yet another kind of performance. If so, they can be considered to share many of the features about which I have already spoken. They are undertaken for the purpose of rendering present something bygone, and aim to tell not just any old story, but to disclose something about a phenomenon: science. They put on display that phenomenon in such a way that certain of its aspects, though possibly already familiar to us, stand out and can be contemplated, lingered over, pondered. Narratives are holistic in that a history is not a catalogue or compendium of one detail after another (which would over-whelm the narrative), but a judicious selection and interweaving of details for the sake of disclosure. Narratives are probative (exploratory) in that one knows not beforehand exactly what will be disclosed when one sets out to construct a narrative, and one allows oneself to be surprised; one is not constructing a narrative when one sets out to find confirming illus-trations of a predetermined thesis. Narratives are provisory in that they are perpetually open to being revised; there is no final narrative about any episode any more than there is a final performance of a play or final experiment in a certain area. Narratives are authoritative in that they de-mand acknowledgment by those engaged in inquiry into the event in ques-tion. They are situational in that they are relative to a certain state of knowledge and perspective; as the perspective or available information

changes, a new narrative may be called for. There is a primacy of performance in narrative; one is not in full control of it, and must put oneself in
the service of the narrative.

The holism of narrative is especially significant. Every detail is poten-
tially revealing. I was once involved, for instance, in a heated discussion
about the disclosive value of candied Mexican hats. In a previous book,
my coauthor and I had related a story of a bet made by a physicist that a
certain particle would be discovered or he would eat his hat. The discovery
was duly made, and at a subsequent conference candied Mexican hats were
passed out for general consumption. A historian of science reproached me
at a conference for devoting space to this episode. What did it contribute
to knowledge about science? Shouldn't I have devoted the space to scientific
information? Hadn't I committed the sin of *popularization*; to focus on extraneous matters because they would be interesting to and comprehensible
by the layperson?

The Mexican hats turned out to be but one instance of a class of details
in my book to which the historian objected. Others included a description
of the handkerchief that students recall Emmy Noether kept in her blouse
and how she waved it when illustrating a point; the flash of an eminent
physicist's florid silk tie as he vanished from students' sight after teaching
a class; the fish that remained uneaten when a brilliant future Nobelist
met his mentor in a restaurant and deferentially allowed the mentor to
order for both of them a dish that the prodigy loathed; the way an Italian
physicist crushed out his cigarette in a film dish; and the comfortable
slippers which a Pakistani physicist working in the West kept underneath
his desk. A professional, so the historian informed me, would have stuck
to the essentials.

I argued in reply that such details properly handled did disclose essential
aspects of science. The bet revealed the game-like quality theoretical physics has for many practitioners. It showed an irreverence for final answers
and rational solutions and a willingness to put oneself on the line; this
quality, in turn, had everything to do with the character of the person who
made the bet and the kind of work that he did. The episode thus served
as an antidote to the view of theorists as solemn fabricators of the ground
plan of the Universe. (The role of comedy and humor in the activity of
science deserves more attention than it has so far received.) Likewise for
the other episodes. The fact that students found Noether's handkerchief
behavior unfeminine indicated the presence of gender stereotypes. The
flash of the tie was emblematic of the obsessive secrecy of the person who
wore it, which in turn was emblematic of the hermetic nature of his work,
which in turn had much to do with the eventual reception of that work in
the scientific community and how little of it was eventually incorporated
into the standard formulations despite the immense achievement it represented. The uneaten fish revealed a mixture of respect and iconoclasm;

that the prodigy was reverential enough to agree to order it on the advice of the mentor but stubborn enough to trust his own taste and refuse to consume it. The film canister/ashtray bespoke the traditional informality and economy of a certain group of Italian scientists. And the slippers were mute testimony of the lonely efforts of a person from the third world to make a home in an unfamiliar environment. Far from serving as mere entertainment, such details were in the service of the disclosure effected by the narrative, and one cannot draw a line between what kinds of details are disclosive and what are not.

It is true that each such detail was inessential in that another, similar one could have been substituted. But that of which the details were disclosive was significant and could not have been omitted; the details were thus symbols. What each detail disclosed could have been made the subject of an explicit study—jokes and gambling in science, sexism, idiosyncracy, mentoring, informality, the anxieties of third-world participants in the international scientific community. Such studies are of course important, but a narrative serves a different function, disclosing a different kind of phenomenon. To object to the inclusion of such details in a narrative has as little justice as to object to the lighting, props, costumes, etc., of a play as having merely entertainment value instead of belonging intrinsically to the performance itself. Indeed, to pass over this kind of detail in narratives about science contributes to the impression that science is a privileged activity unlike other kinds of human activity. Thinking, even scientific thinking, is never conducted in a pure, rarefied environment. Thinking always belongs to the world of appearances, of concrete historical environments. One must beware, therefore, of the impulse to decide beforehand what is irrelevant and what not for a narrative. Any attempt to make such a decision beforehand will be guided by an anticipation of what to expect, by an idea of what the privileged story line is, rather than by the performance itself. To construct a narrative is a process of a back and forth relation between one's ideas about the subject, and what one discovers about it. New anticipations allow us to discover new profiles of the subject, which in turn force us to revise our anticipations.

If narratives about science are akin to performances, then the philosophy of science is akin to the "theory" of the performances. Philosophy of science relies explicitly or implicitly on narratives or accounts of scientific activity, whether extended treatments or anecdotal, and can be thought of as attempting to provide the "theory" of such narrative performances. Too frequently, traditional philosophers of science have relied on mythic or "fictionalized" accounts of science history to support their views.[17] Yet, philosophy of science does not aim to describe an essence above human time and history that works "behind the scenes" of scientific phenomena, but rather to construct a representation of how its characteristic worldly profiles emerge from the processes by which it is produced.

The dialectic between the philosophy of science and narratives about it can be considered analogously to the dialectic between theoretical scripting and experimental performances. Philosophy of science, like theory, allows one to return to the phenomenon—science itself—to look for new profiles and aspects and how they fulfill anticipations. The theatrical analogy, for instance, helps us appreciate aspects that we had not looked at carefully enough before, such as production, recognition, and skill. In highlighting the creative aspect of science, for instance, it might lead one to look for and appreciate expressions of the joy of creation among scientists. The expressions of beauty in Millikan's notebooks, the drunken symposium at the Cosmotron dedication party, the joy of the chase in the *Double Helix*, the satisfaction at knowing about atomic parity violation—all these would then not be particular psychological expressions of individuals but aspects of the practice of science itself insofar as it is a creative and productive worldly activity. True philosophy of science asks questions about science rather than dictates to it, and if things are disclosed about the activity of science it is to allow for new questioning, not to provide things to put up on the shelf as trophies. Other areas that the theatrical analogy opens up for questioning include production, the effect of scale on production, skill, the role of management contracts, the nature of rehearsal/calibration, the nature and character of the laboratory, and the way one can get "swept up" by the theatricality of it all in cases of self-deception.

Moreover, like other kinds of performances, narratives are "produced." That means that someone decides to carry them out, makes necessary decisions in advance, and aims the narrative at a certain community—all of which shape its concrete form. Narratives, too, have many different kinds of research programs. One can look in them for a common theme behind a series of events, look for events having a certain character, and seek out whether there are inconsistencies between a historical event and prevailing characterizations of it.

Viewing narrative as performance thus contributes to a restitution and justification of the storyteller's art. If a narrative is performance and performance disclosure, then the difference between the activity of the subject of the story and the storytelling activity itself does not correspond to that between primordial disclosure and popularization. The storyteller cannot be seen as playing Aaron to Moses. Or, if one insists on putting it that way, it must be with the recognition that their activities are not so fundamentally different because any act of disclosure, even that of Moses, is already a listening. For, as a patient story-listener once reminded me, Moses played Aaron to God.

CONCLUSION
THE PLAY OF NATURE

Is the theatrical analogy really necessary? It is *only* an analogy. But it is only an analogy in the way that taking electromagnetic radiation as waves is only an analogy. Perhaps scientists of the nineteenth century could have started from scratch to develop an account of electromagnetic radiation and begun with something other than wave equations, but what would have been the point? A body of understanding was available that, suitably adapted, revealed its utility for developing an understanding of the phenomenon. The technique of using analogy to develop a philosophical understanding of an obscure phenomenon is an ancient one; what is surprising about the theatrical analogy is that it seems to come from so far afield.

The development of an argumentative analogy involves the adaptation and transformation of the secondary term to bring to bear a single, organized, already articulated perspective on the primary term, and the secondary term then becomes the technically correct one. Electromagnetic radiation, we now say, *is* a wave; the concept of wave was expanded by working out of the analogy, and more kinds of waves are recognized now than before. Similarly, in arguing for the analogy between scientific experimentation and theatrical performance, I am saying that the two kinds of activities are both forms of performance and share a similar structure on an abstract level, and mean to exploit this similar structure to extend our knowledge of the structures of experimentation. This is a book about experiment, not theatre.

The theatrical analogy is thus something to be *used* rather than *proven*. In philosophy, as in other disciplines aiming at the recognition of novel worldly structures, disclosure rather than proof is the culminating event. The reason to invoke theatrical performance argumentatively is to make use of an already developed perspective. At first, the analogy is not descriptive, but after helping to fill in the background, it becomes so. It is part of an *interpretation* of scientific activity, the development of a culturally and historically acquired understanding of it.

One can mean by experimentation either the execution of a particular kind of performance, or a particular kind of inquiry that includes the execution of experimental performances as its principal part. Expressed most concisely, it is the thesis of this book that *experimentation is performance*, with the former term understood in the first of the two senses, and that *science is experimentation,* with the latter term understood in the second of the two senses.

Working out the meaning of performance in the scientific context—making the argumentative analogy itself—is, however, a complex interpretive process. (Thus, in the absence of a critique of the notion of performance, the assertion that science differs from theatre is an illegitimate objection to the program of inquiry outlined here.) The impulse to interpret arises from awareness of a problematic situation; experience of the unsatisfactory character of a present understanding. Working out that interpretation involves three moments. One is a present understanding, or what we *have* now of the problem; a second is a vision of what we lack, or of what a satisfactory account of experimentation might look like; a third is a notion of how to proceed with the reconstruction. Our *Vorhabe* has consisted in part of accounts, however brief, of scientific practice. We have spelled out our *Vorsicht,* or anticipations of what would count as a satisfactory account of experimentation, through the theatrical analogy. In the moment of *Vorgriff* we found ourselves adapting philosophical tools provided by Dewey, Husserl, and Heidegger in effecting the reconstruction. Working out the interpretation creates a new *Vorhabe* or experience of science, in which the schema of pragmatic hermeneutical phenomenology is seen to be fulfilled and for which the theatrical analogy is descriptive rather than metaphorically suggestive.

The vocabulary developed with the aid of the theatrical analogy and Dewey, Husserl, and Heidegger enables us not only to address current issues in the philosophy of science, such as the antinomic character of science, but also to recognize and appreciate features of its practice whose significance is overlooked or dismissed in the traditional account. Experimental results can be murdered and theories massacred without being due to incompetence, malpractice, or error. Scientists can celebrate their work so fiercely that they pass out at tables and set tablecloths afire; they can get sentimental when machines they have worked on for years are turned off. Politics, personalities, management contracts and money are involved in important ways in the planning and execution of experiments. Such things are essential parts of human creative activity of any sort. Does our philosophy of science admit a role for them within a comprehensive picture of science? If it does not, would it not be legitimate to reject it as philosophy of *science* and therefore as *philosophy*?

The study of scientific experimentation is of necessity open-ended. It is impossible to pin down an "essence" of experimentation the way that one can, for instance, pin down the basic and never-changing features of a triangle, so that one can speak from then on with confidence about its past and future forms. (One might even ponder the possibility of a historical *telos* to experimentation.) Experimentation does not aim at a closed, finished structure, nor does it seek to reify or confirm a structure; instead, it seeks continued inquiry. Experimentation is world-building, bringing into the world new beings, though scientists do not thematize this world-building character in their professional activity. Experimentation consists of variations and explorations of the involvements of historically constituted practices with nature under the influence of historically developed theories to disclose different kinds and appearances of phenomena. To think that one could exhaustively describe the nature of experimentation with finality is to presume that one can imagine in advance all of these involvements, practices, theories, and phenomena; it would be like presuming that one can foresee all possible ways in which a play might be performed.

Though open-ended, the examination of experimentation has potential benefits for both science and philosophy. For science, these include both the shedding of light on such puzzles as the understanding of certain features of quantum mechanics as well as the more indirect benefit of allowing an accurate way of suggesting that science, like other human activities, takes place in a social and cultural environment without compromising its due claims to authority. A different set of benefits arises from the fact that science is a human activity always carried out in a cultural and historical context. Its dependence on public funding ensures its dependence on the wider social community; moreover, it affects that wider social community in turn through teaching us how to know our way around in the world, and through the account of the genesis and structure of the world that emerges from science. When the intentions and methods of science are misconceived, the result can be harmful to the general intellectual climate in which scientific activity must be conducted. Consider the growing receptivity to sophistical spiritual doctrines and practices, including astrology, channeling, crystal-gazing, Creationism, and the linkage between ancient Eastern mysticism and quantum mechanics. These doctrines and practices represent partial acceptances of the mythic view of science, as is apparent from their frequent obsessions with precise measurement and contact with purportedly fundamental bodies of nature. But in part, they also represent repudiations of the mythic account of science— an account that, I have argued, ought to be repudiated. However, they want to replace that account with modern variants of ancient traditions that were displaced by modern science precisely because of the latter's growing authority. Because the present-day historical context is entirely different, what such doctrines and practices now involve is largely a recovery of empty vocabularies. There is a social cost to this: while there may be little apparent danger in people reading newspaper horoscopes or purchasing copper bracelets as arthritis cures, habitually doing so makes one receptive to more harmful practices (quack medicine, for instance), the potential damage of which—to both individuals and society at large— should not be underestimated.[1] Moreover, these spiritualistic doctrines and practices are only one manifestation of a more general scientific illiteracy and fear of science, which is far more dangerous to the practice of science than specific cases of opposition to certain scientific projects because of exaggerated fears about what they entail. As physician Gerald Weissman has written, "fear of science is a luxury we cannot afford."[2] The punishment that awaits members of a society that does not have an adequately clarified understanding of the nature of science is to have to live in a society whose individuals—and even some leaders—listen to astrologers.

For philosophy, study of experimentation reveals the pressing need for a notion of phenomenon adequate to the scientific object, although important steps have been made in this direction by Husserl, Heidegger, and Merleau-Ponty. Under the influence of the Göttingen School, Husserl developed an understanding of perception as the appearance of phenomena whose invariance, *more geometrico*, manifested itself through transformations of its profiles. Heidegger and Merleau-Ponty both found this account too abstract and further enriched it, Heidegger by insisting that phenomena do not appear independently of historical and cultural contexts, Merleau-Ponty by emphasizing how phenomena arise out of the interaction between the human body and the world—for the body, not the mind, is the first instrument of inquiry. But these more elaborated concepts still cannot be used to understand scientific phenomena, at least without further modification, and further pursuit of the problem of experimentation will have to involve reworking the notion of phenomenon. Such a reworking may be one of the most important tasks of modern philosophy, given that *phenomenon* is a concept with reverberations for ethics, aesthetics, legal studies, psychoanalysis, and other areas.

The study of experimentation also reveals important clues about the nature and direction of philosophy itself. The neglect of experiment is so thorough and far-reaching that it can hardly have been the result of an oversight. Its neglect is a positive phenomenon that deserves to be examined in its own right for what it reveals about the nature and direction of the philosophical tradition. The neglect of experimentation has been brought about by the mythic view that scientific knowledge is knowledge of the Demiurge, above human history and culture. But no such knowledge is possible, for scientific practice must always remain within the realm of human history and culture and maintain an irreducible tie to human practices even as it discloses phenomena with some degree of independence of any particular historical and cultural context. The problem of experimentation thus involves much more than the relation between experiment and theory, for it brings to light the need not only to question the foundations of the *mythic treatment* of science, but also to question the nature and ambitions of contemporary philosophy. It reveals a deep ontological prejudice according to which reality belongs to the theorist, S_t, or only what is observable from a third-person point of view. The inquiry into experimentation reveals the necessity of further developing a language to articulate the basically artistic experience of the experimenter, S_x.

This prejudice in favor of representation over presentation can be found in theatre as well, where it shows up in the form of taking the script as the principal subject of performance. This is what happens when plays are taught principally as literature, with their performance dismissed as incidental; one recalls Noverre's railings against those who are obsessed with notation in dance at the expense of performance. Artaud's praises of the Balinese theatre, which takes its point of departure from the stage conditions itself, also comes to mind, as well as his condemnations of the Western identification of theatre and text that essentially dismisses everything nontextual as a basically incidental problem of *mise en scène*. "A theatre," he says, "which subordinates the *mise en scène* and production, i.e., everything in itself that is specifically theatrical, to the text, is a theatre of idiots, madmen, inverts, grammarians, grocers, antipoets and positivists, i.e., Occidentals."[3] Heidegger's attacks on the abstractions of science and of theory were made in the same spirit. By using the schema of hermeneutical phenomenology one is not forced into the position of having to choose between presentation or representation; both are accommodated within one framework.

Clearing up confusion about the nature of science helps philosophy in another way. While the confusion mentioned above produces a fear and rejection of science, another kind of confusion produces a worship of it. Science wields enormous influence in government, education, and culture at large, thanks to its tangible fruits and concrete returns; small wonder that it tends to be viewed as the dominant and most successful form of human intellectual inquiry. The mythic view reflects and reinforces this attitude, feeding into the subconscious faith of our culture that science possesses an automatic and inexorable method of gathering information— that it is *the* correct way of answering questions about the world—and encourages belief in the inexorability of scientific progress. But if science is taken to be the paradigm for successful intellectual activity, then other forms of intellectual inquiry—philosophy among them—suffer in contrast. Demythologizing the practice of science is thus necessary to ensure the vitality and authority of philosophy itself. Failure to do so will result in philosophy playing an even more minuscule role in contemporary culture than it does at present, and philosophers will have no one to blame but themselves.

Science, as I said above, can be thought of as experimentation understood as a particular kind of inquiry whose principal feature is the execution of performances. But the forms of scientific inquiry in its various branches are as diverse as its forms of experimentation. Moreover, the process of executing such performances today requires the presence of a large supporting staff of technicians, engineers, educators, materials, information channels, and funding agencies. Science itself, in short, is a culture, one that has been shaped top to bottom by the necessity to conduct various types of performances. Science is routinely portrayed as a culture, most famously in C. P. Snow's book *The Two Cultures*. But Snow and most other commentators say so in order to contrast the culture of science with that of the humanities, and to lament the vast differences and considerable misunderstanding between the two, when what is necessary is to describe and investigate in a rigorous manner the diverse aspects of these cultures. An early and rudimentary attempt to do something like this, which describes different kinds of science for different purposes within a larger community, is Fleck's *Genesis and Development of a Scientific Fact*, written in the 1930s. Fleck viewed science as a profoundly social enterprise and described scientific activity as consisting of highly stratified "thought collectives" that develop over time and which have their own structures. Fleck distinguished between four kinds of science: journal science, vademecum science, textbook science, and popular science. Each of these kinds of science could be investigated in its own right, and has its own institutions, literature, rhetorical styles, audience, purpose, subcultures, and links to other forms. Fleck's book, though, was almost wholly neglected until Kuhn drew attention to it; indeed, Kuhn's own work served to spur historians, sociologists, and philosophers of science to look into the enormous variety of dimensions to the scientific culture.

Developing a *philosophical* account of something with the variety and density of a culture, however, poses a distinct set of problems. Philosophical inquiry does not treat its subject matter as just another species of relevant subject matter, for nothing is in principle outside the scope of philosophy, and what philosophy ultimately seeks of something is its most general character. Not that philosophers have always succeeded in respecting this; the members of the Vienna circle, who had an enduring impact on subsequent philosophy of science, maintained a simplistic and uncritical view of scientific procedure and neglected any ambiguity that it might involve in their promotion of such procedure as *the* standard for intellectual activity of any kind. Even though this approach to science might have been put toward important social ends in the particular historical and cultural milieu of Vienna in the 1920s and 1930s, detached from that functional context, it represents the apotheosis of what Fleck would call popularization. It simplifies scientific activity and glosses over ambiguity and controversy in order to present an (intellectually) exciting program.

A properly philosophical approach to science would insist on looking into the social, historical, and hermeneutical origins of all areas of science, watching them at work in the appearance of phenomena through performances, since this process is responsible for the general character of what science is. A properly philosophical approach would try to find out what happens in those moments when the crucial socio-historical decisions are in the process of being made; it would be fascinated by the presence of ambiguity and controversy and forms of preparation; it would strive to open any "black boxes" or forms of standardization through which accepted and ready-made programs for thinking, acting, and speaking professionally in a given era are passed on. A properly philosophical approach, in other words, would regard its first duty as looking *at the thing itself* to see how it was constituted rather than seeking to put itself in the service of science.

One of the earliest images of the revelation of truth in nature is in Psalm 19: "The heavens declare the glory of God; and the firmament sheweth his handywork." That image—the *epiphany of Nature*—emphasized the visibility of its truths; they are openly disclosed and publicly available to all who would look. However, medieval thinkers such as Paracelsus came to appreciate that special training was required to read that handywork, and those who studied nature and were thus able to reveal his will occupied a growing role as intermediaries between God and the human world.

In the image as revised by Galileo, the truth of nature was depicted less as revealed than as encoded, in the "open book of heaven" that was "written in the language of mathematics." The *book of Nature* image emphasized that the truths of nature were not perspicuous and visible but concealed and difficult to obtain, and implied that searching them out had to be a task for a specially trained professional class of individuals. The image was liberating in Galileo's time in that it directed the attention of inquirers toward the search for invariants of a mathematical character able to be discovered through experimentation. But the image is no longer liberating for us. Relativity and quantum mechanics have called into question the relation between theory and world in a way that challenges the validity of the image.

The picture of nature suggested by development of the theatrical analogy implies yet another image incorporating facets of the first two: the *play of Nature.* Play is here meant in the sense of an infinite, ceaseless activity that exhibits a myriad of forms in as many situations. Yet this play is not chaotic or random but governed by patterned, discoverable constraints. Thus the image retains the open, revealed character of the activity—its accessibility to perception—but at the same time its scriptedness. This play transpires for the most part without human participation, but human beings can and do participate through interventions in order to supplement their awareness of the various forms of play, and furnish them with material for scriptwriting. But as their perceptions of the play become developed, deepened, and enriched, the scripts change accordingly.

At the beginning of this book, I cited the remarks of three prominent scientists who were critical of the philosophical attempt to understand the nature of what they do. These remarks are typical of others that one hears when one reveals to working scientists that one is a philosopher of science;

generally, the remarks are politely dismissive and mildly patronizing. I have shown why such accusations have a surface plausibility, for existing philosophy of science has ignored many features of science, overemphasized others, and in general succumbed to a mythic view that it itself has largely constructed, maintained, and promoted. It has also sought to enlist philosophy in championing and participating in the scientific enterprise. One of the formative figures of traditional philosophy of science, Hans Reichenbach, once wrote that modern science has become so complicated that the modern scientist has lost track of the philosophical implications of theories: "He then became aware that he was walking, so to speak, on the thin ice of a superficially frozen lake, and he realized that he might slip and break through at any moment."[4] At this point, according to Reichenbach, the philosopher steps in to the rescue—not only of the particular scientist, but of science itself.

Genuine philosophy instinctively recoils from such presumption. It questions received images and probes its objects in the light of its own traditions; it does not put itself at the service of what it interprets. To return to the analogy of the bookshelves elaborated in the second chapter, philosophy creates shelves that address some subject in the world, and does so only for the sake of seeing it better. It seeks to multiply the perspectives on its objects for its own sake. Assuming that the world is already disclosed, philosophy has no need of experimental performances or syntactical models to represent such performances; its performances consist of the process of assembling the shelves, and of making the books on it "talk" to one another about the subject matter in the world in more or less the ordinary semantical vocabulary.

Finally, however, there is no a priori defense against the accusations of the kind mentioned at the beginning of this book, except to show in deed, in *ergon*, what philosophy of science does. For apart from a surface plausibility, the accusations also have a *deep* plausibility as well, because philosophy inevitably lends itself to this kind of misunderstanding. Its aim is to multiply perspectives on phenomena in the world for its own sake, rather than to see how they might be applied to particular ends in the world; a philosopher temperamentally puts something in doubt rather than solves a problem, and prefers to dredge up assumptions and presuppositions that in the natural attitude would remain hidden. Inevitably, someone intent upon accomplishing existing tasks will find such questioning distracting, subversive, and corrupting on the one hand, and trivial and foolish on the other. Sophism, the mere pretense to wisdom, is an occupational hazard of philosophy, and its spectre always hangs over the field. The only defense that one has for not having fallen victim to it lies in the intelligibility that one is able to generate. When the work of philosophers is able to provide that intelligibility, then it rejoins the ancient meaning of apo-logia, to speak from the *logos*.

NOTES

Introduction

1. Weinberg's remark is from Steven Weinberg, "The Forces of Nature," *American Scientist* 65 (1977): 175; Gell-Mann's is cited in Robert P. Crease, "Good Philosophy and Good Physics," in the *Threepenny Review* 15 (Fall 1983): 9, from an interview with the author; Feynman's is from Richard P. Feynman, *Surely You're Joking, Mr. Feynman!* (New York: Norton, 1985), p. 70. See also Steven Weinberg, *Dreams of a Final Theory* (New York: Pantheon, 1992), chapt. 7, "Against Philosophy," which is, however, an amateur philosophical rumination itself. For the "harsher allegations," see below.

2. Hans Reichenbach, for instance, writes that "The philosopher of science is not much interested in the thought processes which lead to scientific discoveries; he looks for a logical analysis of the completed theory." In "The Philosophical Significance of the Theory of Relativity," *Albert Einstein: Philosopher-Scientist,* ed. Paul A. Schilpp (Evanston, Ill.: Library of Living Philosophers, 1970), p. 292. See also Karl Popper's distinction between the *logic* of knowledge—what he sees as the aim of the authentic philosophy of science—and the *psychology* of knowledge. Karl R. Popper, *The Logic of Scientific Discovery* (New York: Harper and Row, 1968), p. 31. The impact of Reichenbach's distinction between "context of justification" and "context of discovery" is to release the philosopher of science from the burden of having to examine actual practice.

3. Immanuel Kant, *Critique of Pure Reason,* trans. Norman K. Smith (New York: St. Martin's Press, 1965), pp. 20–21 (B xiii–xiv).

4. Paul Edwards, ed., *The Encyclopaedia of Philosophy* (New York: Macmillan, 1972); Morris R. Cohen, *Reason and Nature: The Meaning of Scientific Method* (London: Free Press, 1964); John Earman, ed., *Testing Scientific Theories,* Minnesota Studies in the Philosophy of Science vol. 10 (Minneapolis: University of Minnesota Press, 1983); Ernest Nagel, *The Structure of Science* (New York: Harcourt, Brace, 1961), p. 15.

5. The *hoti,* literally the "that" of a thing, has been translated as the "facts" about it. This translation is misleading at best, since after the *hoti* are collected and examined they may turn out to have been misconceptions and need to be readjusted.

6. See: Robert Ackermann, *Data, Instruments, and Theory* (Princeton: Princeton University Press, 1985); Allan Franklin, *The Neglect of Experiment* (New York: Cambridge University Press, 1986); Peter Galison, *How Experiments End* (Chicago: University of Chicago Press, 1987); David Gooding, *Experimentation and the Making of Meaning* (New York: Cambridge University Press, 1990); Trever Pinch, David Gooding, and Simon Schaffer, eds., *The Uses of Experiment* (New York: Cambridge University Press, 1989); Ian Hacking, *Representing and Intervening* (New York: Cambridge University Press, 1983); Patrick A. Heelan, *Space-Perception and the Philosophy of Science* (Berkeley: University of California Press, 1983); Don Ihde, *Technics and Praxis* (Boston: Reidel, 1979); Andrew Pickering, *Constructing Quarks* (Chicago: University of Chicago Press, 1984); Joseph Rouse, *Knowledge and Power: Toward a Political Philosophy of Science* (Ithaca: Cornell University Press, 1987); Steven Shapin and Simon Schaffer, *Leviathan and the Air-Pump: Hobbes, Boyle, and the Experimental Life* (Princeton: Princeton University Press, 1985).

7. "It is another property of the human mind that whenever men can form no idea of distant and unknown things, they judge them by what is familiar and at hand." In Giambattista Vico, *The New Science,* trans. Thomas G. Bergin and Max H. Fisch (Ithaca: Cornell University Press, 1968), p. 60. Jeremy Bernstein, *Elementary Particles and Their Currents* (San Francisco: W. H. Freeman, 1968), p. vii.

8. Gilbert N. Lewis, *The Anatomy of Science* (New Haven: Yale University Press, 1926), p. 1.

9. William Broad and Nicholas Wade, *Betrayers of the Truth: Fraud and Deceit in the Halls of Science* (New York: Simon and Schuster, 1982), p. 139.

10. Martin Gardner, *Fads and Fallacies in the Name of Science* (New York: Dover, 1957), p. 332. For another example of experiments gone awry, see the episode involving the establishment of the Universal Fermi Interaction in Robert P. Crease and Charles C. Mann, *The Second Creation: Makers of the Revolution in 20th Century Physics* (New York: Macmillan, 1986), chapt. 11.

11. Look for an increase of this desire in the wake of the David Baltimore case and other recent widely publicized instances of fraud.

12. Adam Smith, "Essays in Philosophical Subjects," *The Early Writings of Adam Smith,* ed. J. Ralph Lindgren (New York: Augustus M. Kelley, 1967), p. 49.

13. Boris Hessen, "The Social and Economic Roots of Newton's *Principia,*" *Science at the Crossroads,* ed. D. Hymer (London: Frank Cass, 1971): 149–212; Paul Forman, "Weimar Culture, Causality and the Quantum Theory, 1918–1927: Adaptation by German Physicists and Mathematicians to a Hostile Cultural Environment," *Historical Studies in the Physical Sciences* (1971): 1–115; and "*Kausalität, Anschaulichkeit, and Individualität,* or How Cultural Values Prescribed the Character and the Lessons Ascribed to Quantum Mechanics," in *Society and Knowledge,* eds. Nico Stehr and Volker Meja, (New Brunswick, N.J.: Transaction Books, 1984): 333–47.

14. Some of the more important works in the tradition of social constructivism include: Bruno Latour and Steve Woolgar, *Laboratory Life: The Social Construction of Scientific Facts* (London: Sage, 1979); Karin D. Knorr-Cetina, *The Manufacture of Knowledge: An Essay on the Constructivist and Contextual Nature of Science* (Oxford: Pergamon, 1981); Bruno Latour, *Science in Action* (Cambridge: Harvard University Press, 1987). For an important critique from the perspective of a sociologist, see Stephen Cole, *Making Science: Between Nature and Society* (Cambridge: Harvard University Press, 1992).

15. Margaret Jacob, *The Cultural Meaning of the Scientific Revolution* (New York: Knopf, 1988), p. 86.

16. T. Theocharis and M. Psimopoudos, "Where Science Has Gone Wrong," *Nature* 329 (1987): 595–98.

17. As was the conclusion of a 1991 meeting of the British Association for the Advancement of Science on the 150th anniversary of the name "dinosaur."

18. Graham's original remarks first appeared in *This I Believe,* ed. Edward P. Morgan (London: Hamish Hamilton, 1953), pp. 136–37, but were revised in 1987—and the reference to science added—when they were included in the program printed for a 1987 appearance of her company in New York City.

19. Theocharis and Psimopoudos, "Where Science Has Gone Wrong," p. 595.

1. The Mythic Account of Experimentation

1. See, for instance, Steve Fuller, *The Philosophy of Science and Its Discontents* (Boulder: Westview Press, 1989). Hacking refers to a "crisis of rationality" in his *Representing and Intervening,* p. 1.

2. "Democritian" is slightly misleading, because Democritian atoms are geometrical.

3. In his entry, "Galilei, Galileo," in the *Dictionary of Scientific Biography*, Stillman Drake says that the alleged demonstration at the Leaning Tower, if actually performed, "was clearly not an experiment but a public challenge to the philosophers." *Dictionary of Scientific Biography*, ed. Charles C. Gillespie (New York: Scribner's, 1972), p. 238.

4. See, for instance, Lane Cooper, *Aristotle, Galileo and the Tower of Pisa* (Ithaca: Cornell University Press, 1935), though this account is controversial.

5. Carl G. Adler and Byron L. Coulter, "Galileo and the Tower of Pisa Experiment," *American Journal of Physics* 46 (1978): 199–201. Franklin, *Neglect of Experiment*, pp. 1f., contains a brief discussion of the mythic aspects of this experiment.

6. Galileo, *Discoveries and Opinions of Galileo*, trans. Stillman Drake (New York: Doubleday, 1957), pp. 196, 238.

7. See, for instance, Arthur S. Eddington, *The Nature of the Physical World* (New York: Macmillan, 1948), pp. ix–xvii, and Wilfrid Sellars, "Philosophy and the Scientific Image of Man," *Science, Perception and Reality* (New York: Humanities Press, 1971), pp. 1–40.

8. For a commentary on this mythic view of science see Ian Mitroff, *The Subjective Side of Science: A Philosophical Inquiry into the Psychology of the Apollo Moon Scientists* (New York: Elsevier, 1974). See also Stephen G. Brush's playfully titled article, "Should the History of Science Be Rated X?" in *Science* 183 (1974): 1164–72.

9. Albert Einstein, *The World As I See It* (New York: Friede, 1934), p. 22.

10. Richard Feynman, Robert B. Leighton, and Matthew Sands, *The Feynman Lectures on Physics*, vol. 1 (Menlo Park: Addison-Wesley, 1963), p. 2-1.

11. Stephen W. Hawking, *A Brief History of Time* (New York: Bantam, 1988), p. 10. Three pages later he writes, "our goal is nothing less than a complete description of the universe we live in."

12. Albert Einstein, "On Scientific Truth," in *Ideas and Opinions*, ed. Carl Seelig (New York: Dell, 1976), p. 255.

13. Hawking, *Brief History*, p. 175.

14. Allan Franklin, *The Neglect of Experiment* (New York: Cambridge University Press, 1986), pp. 1 and 3; the quotes below are from pp. 3 and 244. See also Allan Franklin, *Experiment, Right or Wrong* (New York: Cambridge University Press, 1990).

15. The critical edition of the works of John Dewey are issued in three series, *The Early Works (1882–1898)*, *The Middle Works (1899–1924)*, and *The Later Works (1925–1953)*, by Southern Illinois University Press. I shall follow the standard system and cite, immediately following the quote, the initials of the series followed by volume and page number. *The Quest for Certainty* is LW4. This reference is to LW4:6.

16. "Thought experiments," which require no equipment other than the imagination, are only an apparent exception; rather than exploring nature, they test the consistency of a theory. For some studies on thought experiments, see Tamara Horowitz and Gerald Massey, eds., *Thought Experiments in Science and Philosophy* (Savage, Md.: Rowman and Littlefield, 1991).

17. Rene Descartes, *Regulae ad directionem ingenii*, ed. and trans. G. Le Roy (Paris: Boivin: 1933).

18. Bertrand Russell, *Our Knowledge of the External World, as a Field for Scientific Method in Philosophy* (London: G. Allen and Unwin, 1926); Ludwig Wittgenstein, *Tractatus Logico-Philosophicus*, trans. D. P. Pears and B. F. McGuinnes (Atlantic Highlands, N.J.: Humanities Press, 1974).

19. Otto Neurath, "Physicalism: The Philosophy of the Viennese Circle," *The Monist* 41 (1931): 619; *The Scientific Conception of the World: The Vienna Circle*, which was originally a pamphlet with no author on the title page, but was written by

Otto Neurath and edited by Rudolf Carnap and Hans Hahn. It was published as an article in Otto Neurath, *Empiricism and Sociology*, ed. Marie Neurath and Robert S. Cohen (Boston: Reidel, 1973), p. 306. See also Thomas E. Uebel, ed., *Rediscovering the Forgotten Vienna Circle* (Boston: Kluwer, 1991).

20. Carnap writes that while in simple observation we merely look at things like flowers and trees and try to discern regularities, in experiments we are no longer simply onlookers but "we *do* something that will produce better observational results than those we find by merely looking at nature." In Rudolf Carnap, *Philosophical Foundations of Physics* (New York: Basic Books, 1966), p. 40. The insight implied in the remark that experimentation is a doing is not exploited.

21. Rudolf Carnap, "Protocol Statements and the Formal Mode of Speech," in *Essential Readings in Logical Positivism*, ed. Oswald Hanfling (Oxford: Blackwell, 1981), p. 153.

22. In a conversation I once had about the philosophy of science with the late Alfred J. Ayer, a member of the Vienna Circle, the latter not only professed his complete ignorance of experimental techniques and problems but also his utter lack of interest in the subject and conviction that it had nothing to do with the philosophy of science.

23. Sellars, "Philosophy," p. 173.

24. Hacking, *Representing and Intervening*. While in Part A of his book Hacking comes down on the side of a variety of realism, in Part B he begins to develop the insight that experiments do not merely check but also create phenomena, and it is never completely clear how the two positions are to be adequately reconciled. The point of the theatrical analogy will be to suggest a model for just this kind of reconciliation.

25. Carnap wrote of abandoning his previous belief in a "rock bottom of knowledge, the knowledge of the immediately given, which was indubitable." In *The Philosophy of Rudolf Carnap*, ed. Paul A. Schilpp, (La Salle, Ind.: Open Court, 1963), p. 57.

26. Karl Popper, *The Logic of Scientific Discovery* (New York: Harper and Row, 1968), p. 111.

27. Ibid., p. 107, n. 3.

28. See, for instance, Quine's classic essay "Two Dogmas of Empiricism," in *From a Logical Point of View* (Cambridge: Harvard University Press, 1980), pp. 20–46; and W. V. O. Quine and J. S. Ullian, *The Web of Belief* (New York: Random House, 1978).

29. C. P. Snow, *The Two Cultures and a Second Look* (New York: Cambridge University Press, 1982).

30. Albert Einstein, *Out of My Later Years* (New York: Philosophical Library, 1950), p. 59.

31. It may be a flaw of this book that too much attention is paid to one particular branch of science, namely physics, rather than evenly distributed among the various branches. However, this is not due to an a priori assumption that physics is the model science but rather to the background of the author.

32. Immanuel Kant, *Critique*, p. 539 (A 653, B 681).

33. From *The Innocents Abroad* (New York: Library of America, 1984), pp. 196–97.

2. Philosophers and Productive Inquiry

1. I owe this wonderful metaphor to Patrick A. Heelan, "Hermeneutical Philosophy and the History of Science," in *Nature and Scientific Method: William A. Wallace Festschrift*, ed. Daniel O. Dahlstrom (Washington, D.C.: Catholic University of America Press, 1990), p. 27. Inevitably, the use I make of it differs somewhat from Heelan's.

2. See, for instance, Robert P. Crease and Charles C. Mann, "The Yogi and the Quantum," in *Philosophy of Science and the Occult*, 2nd ed., ed. Patrick Grim (Albany: State University of New York Press, 1990.

3. And even if they are not, for philosophy has its share of what Gore Vidal calls "scholar-squirrels."

4. Recent works relevant to Dewey and science include Larry A. Hickman, *John Dewey's Pragmatic Technology* (Bloomington: Indiana University Press, 1989); and R. W. Sleeper, *The Necessity of Pragmatism* (New Haven: Yale University Press, 1986), esp. chapt. 7.

5. "Experience, the actual experience of men, is one of doing acts, performing operations, cutting, marking off, dividing up, extending, piecing together, joining, assembling and mixing, hoarding and dealing out; in general, selecting and adjusting things as means for reaching consequences. Only the peculiar hypnotic effect exercised by exclusive preoccupation with knowledge could have led thinkers to identify experience with reception of sensations" (LW4:125).

6. Quoted in *SLAC Beam Line* (the bulletin of the Stanford Linear Accelerator Facility, in Stanford, Calif.) 19 (September 1989): 1.

7. Hickman, *Dewey's Pragmatic Technology*, chapt. 2.

8. Ibid., p. 99.

9. Alan Lightman and Owen Gingerich, "When Do Anomalies Begin?" *Science* 255 (1992): 690–95.

10. Arthur S. Eddington, *The Nature of the Physical World* (New York: Macmillan, 1948), p. ix.

11. For a more extended and sympathetic characterization of Dewey's account of scientific entities, see Sleeper, *Necessity of Pragmatism*, chapt. 6.

12. On Husserl's philosophy of science, see Patrick A. Heelan, "Husserl's Later Philosophy of Science," *Philosophy of Science* 54 (1987): 368–90; "Husserl, Hilbert, and the Critique of Galilean Science," in *Edmund Husserl and the Phenomenological Tradition*, ed. Robert Sokolowski (Washington, D.C.: Catholic University of America Press, 1988): 157–73; and "Husserl's Philosophy of Science," in *Husserl's Phenomenology: A Textbook*, eds. Jitendranath N. Mohanty and William R. McKenna (Washington, D.C.: University Press of America and The Center for Advanced Research in Phenomenology, 1989): 387–427, Theodore Kisiel, "Phenomenology as the Science of Science," in *Phenomenology and the Natural Sciences*, eds. Joseph J. Kockelmans and Theodore Kisiel, (Evanston: Northwestern University Press, 1970): 5–44; and "Husserl on the History of Science," in ibid., pp. 68–92, Joseph Kockelmans, "The Mathematization of Nature in Husserl's Last Publication," in ibid., pp. 45–67.

13. For a bibliography on phenomenological approaches to natural science, see Steven Chasan, "Bibliography of Phenomenological Philosophy of Natural Science," in *Phenomenology of Natural Science*, eds. Lee Hardy and Lester Embree (Washington, D.C.: University Press of America and The Center for Advanced Research in Phenomenology, 1992).

14. Edmund Husserl, *Ideas Pertaining to a Pure Phenomenology and to a Phenomenological Philosophy, First Book*, trans. Fred Kersten (Boston: Nijhoff, 1983), p. 35.

15. Edmund Husserl, *Cartesian Meditations*, trans. Dorion Cairns (The Hague: Nijhoff, 1973), p. 45.

16. For an elaboration of this point, see Maurice Merleau-Ponty, *Phenomenology of Perception*, trans. Colin Smith (London: Routledge and Kegan Paul, 1962), part 1, chapt. 2; and part 2, chapt. 3.

17. Edmund Husserl, *The Crisis of European Sciences and Transcendental Phenomenology*, trans. David Carr (Evanston: Northwestern University Press, 1970), p. 166.

18. See David Carr, "Husserl's Problematic Concept of the Life-World," in *Hus-*

serl: Expositions and Appraisals, eds. Frederick A. Elliston and Peter McCormick (Notre Dame: University of Notre Dame Press, 1977): 202–12.

19. The page numbers of subsequent references to this work will be identified in parentheses immediately following the quote.

20. See David Carr, "Husserl's Problematic."

21. For the phrase "cultural niche," see Patrick A. Heelan, "After Experiment: Realism and Research," in *American Philosophical Quarterly* 26 (1989): 297–308.

22. Abraham Pais, *Inward Bound* (New York: Oxford University Press, 1986), p. 67.

23. An immense secondary literature exists on Heidegger that would be pointless to summarize here; for a survey of the general trends of Heidegger scholarship in this country, see David Kolb, "Heidegger at 100, in America," *Journal of the History of Ideas* 52 (1991): 140–51.

24. Martin Heidegger, *Being and Time,* trans. John Macquarrie and Edward Robinson (New York: Harper and Row, 1962), sections 15–18. The page numbers of subsequent references to this work will be listed in parentheses immediately following the quote.

25. For the term "flesh" here, see Patrick A. Heelan, "Yes! There Is a Hermeneutics of Natural Science: A Rejoinder to Markus," in *Science in Context* 3 (1989): 477–88; Heelan borrows the term in turn from Maurice Merleau-Ponty, *The Visible and the Invisible,* trans. Alphonso Lingis (Evanston: Northwestern University Press, 1968), eg. p. 127.

26. Martin Heidegger, *Vom Wesen des Grundes,* 3rd ed. (Frankfurt: Klostermann, 1949), p. 15.

27. Magda King, *Heidegger's Philosophy* (New York: Macmillan, 1964), pp. 6–7.

28. Thomas S. Kuhn, *The Essential Tension: Selected Studies in Scientific Tradition and Change* (Chicago: University of Chicago Press, 1977), chapts. 2 and 10.

29. Cf. section 44 of *Being and Time.*

30. This distinction, though based in Heidegger, was not made explicitly by him. But it has been made by others; see, for instance, Heelan, "A Rejoinder to Markus." Heelan's distinction is between "weak" or what I call text hermeneutics and "strong" or what I call act hermeneutics. I find this terminology unsatisfactory, however, insofar as the two versions of the hermeneutical circle involved are on a par from the point of view of the forestructures of understanding; one is not a more rigorous, committed, or risky version of the other.

31. For an account of this episode, see Crease and Mann, *Second Creation,* chapt. 11.

32. But see also sections 13 and 15–21.

33. "When the basic concepts of that understanding of Being by which we are guided have been worked out, the clues of its methods, the structure of its way of conceiving things, the possibility of truth and certainty which belongs to it, the ways in which things get grounded or proved, the mode in which it is binding for us, and the way it is communicated—all these will be determined. The totality of these items constitutes the full existential conception of science" (414).

34. Compare numerous popular images of science, such as those expressed in the movies *E.T.* and *Splash!* in which the cold, unfeeling scientists nearly kill the vital but somehow also defenseless protagonists, or that expressed by Poe in the sonnet "To Science," in which the discipline is a "vulture" which has driven deities from the wood and that preys upon the poet's heart.

35. Martin Heidegger, *The Question Concerning Technology* (New York: Garland, 1977), pp. 121–22.

36. Ibid., p. 27. One can, of course, amuse oneself by catching scientists making

comments that seem to illustrate the truth of Heidegger's remark. A *New York Times* story of September 25, 1990, for instance, reports an engineer who was working on a device to tap energy from the waves striking the rocky western coast of Scotland as saying, "You watch that tremendous power all the way down the coast, and it makes you think of the hundreds of megawatts that are being dashed on the shore and not being used. The ocean is like a big battery, a huge collector and we can tap it in many places." But whether this remark is representative of all scientists, or illustrates a "scientific attitude," or even sheds any light on the way that that particular scientist actually went about planning, building, and operating the device, are entirely different questions.

37. Martin Heidegger, *Discourse on Thinking*, trans. John M. Anderson and E. Hans Freund (New York: Harper and Row, 1966), p. 50.

38. Martin Heidegger, "Postscript" to "What is Metaphysics?" trans. R. F. C. Hull and Alan Crick, in *Martin Heidegger: Existence and Being*, ed. Werner Brock (South Bend, Ind.: Gateway, 1949, p. 353.

39. Maurice Merleau-Ponty, "Eye and Mind," in *The Essential Writings of Merleau-Ponty*, ed. Alden L. Fisher (New York: Harcourt, Brace and World, 1969), p. 252; *Phenomenology of Perception*, trans. Colin Smith (New Jersey: Humanities Press, 1962), p. ix.

40. Hans-Georg Gadamer, *Truth and Method* (New York: Seabury Press, 1975), p. 409.

41. Ibid., p. 412.

42. See, for instance, Max Horkheimer and Theodor Adorno, *Dialectic of Enlightenment* (New York: Seabury Press, 1979).

43. See, for instance, Joseph Rouse, *Knowledge and Power*. Rouse uses tools of the hermeneutical tradition to develop a sophisticated account of how scientific practices operate in, shape, and disclose entities in the scientific world, but fails to take advantage of tools provided by the phenomenological tradition to provide an account of what it is that is so disclosed. Rouse is content to say that the question of what is disclosed is "an entirely objective matter" (159). However, it is critical to discover the nature of that objectivity, for astrology, witchcraft, and various pseudodiscoveries also had their own practices, their own interpretations, their own power relations, their own standardizations, their own world disclosed in the practices. While rightly insisting that science is an activity that discloses entities in the world, Rouse wrongly neglects the question of the being of the entities so disclosed. The small role played by theory in his account is symptomatic of this neglect; theories are best viewed, he says, as "strategies for dealing with various phenomena" (116). Rouse is evidently afraid that granting a large role to theory would force him to surrender the insights of hermeneutics and the disclosive effects of power relations. In effect, what Rouse has done is taken the traditional priority of theory over praxis and stood it on its head, when what is needed is a rethinking of that relation.

3. Experimentation as a Performing Art

1. Aristotle, *Poetics* 1457b, trans. Ingram Bywater, in *The Basic Works of Aristotle*, ed. Richard McKeon (New York: Random House, 1941), at p. 1476, slightly altered.

2. Max Black, *Models and Metaphors* (Ithaca: Cornell University Press, 1962), pp. 39–41. See also, Max Black, "More about Metaphor," *Dialectica* 31 (1977): 431–57.

3. "Man is not . . . a wolf." Jean-Jacques Rousseau, "Letter to M. D'Alembert on the Theatre," in *Politics and the Arts*, trans. Allan Bloom (Ithaca: Cornell University Press, 1960), p. 86.

4. Black, *Models*, p. 41.

5. Bruce Wilshire, *Role Playing and Identity: The Limits of Theatre as Metaphor* (Bloomington: Indiana University Press, 1982), p. 94.

6. See Introduction, footnote 7.

7. "It is another property of the human mind that whenever men can form no idea of distant and unknown things, they judge them by what is familiar and at hand." Giambattista Vico, *The New Science*, trans. Thomas G. Bergin and Max H. Fisch (Ithaca: Cornell University Press, 1968), p. 60.

8. From a news item with no title and no byline, *Science News* 137 (1990): 359. The person quoted is Berlin University astronomer Michael Mendillo.

9. Robert Millikan, *Autobiography* (New York: Prentice-Hall, 1950), p. 80.

10. Cited in Evelyn Fox Keller, *Reflections on Gender and Science* (New Haven: Yale University Press), 1985, p. 165.

11. Heelan, *Space-Perception*; "Hermeneutical Philosophy"; "After Experiment"; "Husserl's Philosophy of Science"; "Experiment and Theory: Constitution and Reality," *Journal of Philosophy* 85 (1988): 515–24; "Husserl, Hilbert, and the Critique of Galilean Science"; "Husserl's Later Philosophy of Science," *Philosophy of Science* 54 (1987): 368–90; "Machine Perception," in *Philosophy and Technology II*, eds. C. Mitcham and A. Huning (Boston: Reidel, 1986): 131–56; "Natural Science as a Hermeneutic of Instrumentation," *Philosophy of Science* 50 (1983): 181–204; "Perception as a Hermeneutical Act," *Review of Metaphysics* 37 (1983): 61–75; "Natural Science and Being-in-the-World," *Man and World* 16 (1983): 207–19. Unless otherwise specified, page numbers of subsequent references to Heelan will be from *Space-Perception*, and identified in parentheses immediately following the quote.

12. Rouse, *Knowledge and Power*, p. 146.

13. See, for instance, Mary Jo Nye, *Science in the Provinces: Scientific Communities and Provincial Leadership in France, 1860–1930* (Berkeley: University of California Press, 1986), chap. 2.

14. For an account, see Crease and Mann, *Second Creation*, chaps. 17 and 18.

15. Heelan, for instance, uses Dirac state vector notation to call the model state vector $|X>_t$ and the empirical or phenomenological state vector $|X>_x$ (the ket vector is a set-theoretic, abstract mathematical object representing the state of the system, corresponding to Husserl's objects formed of *Mannigfältigkeiten*. Heelan writes, "The theoretical model $|X>_t$ now appears from the standpoint of S_t to be a kind of language under which the reference phenomenon with its empirical profiles $|X>_x$ is described. . . . The syntax of that 'language' is mathematical; of itself it provides no more than what Husserl would have called a *formal ontology* or a *possible* formal ontology, that is, empty schemata of categories of things. Its semantics, however, is tied up with the standardized experimental praxis within which it is used." Heelan, "Experiment and Theory", p. 521.

16. Heelan, "Experiment and Theory," p. 522.

17. "The last word in scientific research is then with S_x." Heelan, "Experiment and Theory", p. 522.

18. Heelan, "After Experiment", p. 303.

19. Ibid., p. 297.

20. Rudolf Carnap, "Empiricism, Semantics and Ontology," in *Semantics and the Philosophy of Language*, ed. Leonard Linsky (Urbana: University of Illinois Press, 1966): 216–21. For a critique of such "prescriptivism," see David Weissman, *Intuition and Ideality* (Albany: State University of New York Press, 1987), and *Hypothesis and the Spiral of Reflection* (Albany: State University of New York Press, 1989).

21. W. V. O. Quine, "Two Dogmas of Empiricism," in *From a Logical Point of View*, p. 43.

22. Hilary Putnam, *Reason, Truth and History* (Cambridge: Cambridge University Press, 1981), p. 52.

23. Hilary Putnam, *Meaning and the Moral Sciences* (London: Routledge and Kegan Paul, 1978), p. 184.

24. See, for instance, David Gooding, "How Do Scientists Reach Agreement About Novel Observations?" *Studies in History and Philosophy of Science* 17 (1986): 205–30, and *Experimentation and the Making of Meaning* (New York: Cambridge University Press, 1990); Thomas Nickles, "Justification and Experiment," in *The Uses of Experiment*, eds. David Gooding, Trevor Pinch, and Simon Schaffer (New York: Cambridge University Press, 1989): 299–333.

25. "Experiment and Theory," p. 516.

26. For a discussion of the use and abuse of theatre as metaphor, see Wilshire, *Role Playing and Identity.*

27. Thus the difference between this use of the theatrical metaphor and that of Arthur Fine, who in *The Shaky Game: Einstein, Realism, and the Quantum Theory* (Chicago: University of Chicago Press, 1986), p. 148, refers to science as "a sort of grand performance, a play or opera, whose production requires interpretation and direction"; moreover, "audience and crew play as well."

28. Aristotle had spoken of *episteme theoretike*, paying attention to eternal and unchangeable things, and naturalizations of *theoria* in this context gave it some currency as referring to observations concerning the first principles of nature. Around the seventeenth century, however, the term also acquired the meaning of the thinking of first principles of nature *as opposed to* the experiences or observations that established, confirmed, or manifested them. If science meant the inquiry into nature, theory now meant the state of our knowledge about its foundations. Theory thus addressed the foundations of nature and was disengaged from them, as a blueprint from a completed building. Hilbert, in the second decade of this century, provided what might be called the modern formulation of the meaning of a scientific theory; the theory of a field of knowledge is a deductive mathematical system based on a few fundamental axioms that provides an ontology for that field and also specifies the ideal limit of measurements.

29. Francis Bacon, *Novum Organum*, #44, in *Great Books of the Western World*, vol. 30 (Chicago: Encyclopaedia Britannica, 1952), p. 110.

30. The exceptions include Bruce Wilshire, *Role Playing and Identity.*

31. Ibid., p. 97.

32. James Watson, *The Double Helix* (New York: Norton, 1980), p. 25.

33. John Caputo, *Radical Hermeneutics* (Bloomington: Indiana University Press, 1987), p. 37.

34. Ludwik Fleck, *Genesis and Development of a Scientific Fact*, trans. Fred Bradley and Thaddeus J. Trenn (Chicago: University of Chicago, 1979), chapt. 3.

4. Performance: Presentation

1. See, for instance, George Herbert Mead, *The Philosophy of the Act* (Chicago: University of Chicago Press, 1938). Mead distinguishes four basic stages of action: impulse, perception, manipulation, and consummation.

2. For a bibliography of general literature on laboratories, see Frank A. J. L. James, *The Development of the Laboratory* (New York: American Institute of Physics, 1989). See also K. Everett and D. Hughes, *A Guide to Laboratory Design* (Boston: Butterworths, 1975).

3. For "space of action" see Elisabeth Ströker, *Investigations in Philosophy of Space*, trans. Algis Mickunas (Athens: Ohio University Press, 1987), p. 48.

4. See, for instance, Robert P. Crease, "The History of Brookhaven National

Laboratory, Part One: the Graphite Reactor and the Cosmotron," *Long Island Histori-
cal Journal* 3 (1991): 167–88; John L. Heilbron and Robert W. Seidel, *Lawrence and His
Laboratory: A History of the Lawrence Berkeley Laboratory,* vol. 1 (Berkeley: University of
California Press, 1989); Lillian Hoddeson, "Establishing KEK in Japan and Fermilab
in the US: Internationalism, Nationalism, and High Energy Accelerators." *Social
Studies of Science* 13 (1983): 1–48.

5. Carol R. Clemmer and Thomas P. Beebe, Jr., "Graphite: A Mimic for DNA
and Other Biomolecules in Scanning Tunneling Microscope Studies," *Science* 251
(1991): 640–42.

6. Gerald Holton, *The Scientific Imagination* (New York: Cambridge University
Press, 1978), esp. chapt. 2, and *Thematic Origins of Scientific Thought* (Cambridge:
Harvard University Press, 1973).

7. See Michael Polanyi, *Personal Knowledge: Towards a Post-Critical Philosophy*
(Chicago: University of Chicago Press, 1962), and *The Tacit Dimension* (Gloucester,
Mass.: Peter Smith, 1983).

8. Evelyn Fox Keller, *Reflections on Gender and Science* (New Haven: Yale University Press, 1985), esp. chapt. 6.

9. See, for instance, Nicholas Rescher, ed., *Aesthetic Factors in Natural Science*
(New York: University Press of America, 1990).

10. From Gottlob Frege, "On Sense and Reference," in *Translations From the Philosophical Writings of Gottlob Frege,* eds. Peter Geach and Max Black (Oxford: Blackwell,
1952), p. 63.

11. Fleck, *Genesis and Development,* chapt. 3.

12. From Crease and Mann, *Second Creation,* pp. 337–38.

13. This language is drawn from Caputo, *Radical Hermeneutics,* esp. chapt. 2.

14. Gadamer, *Truth and Method,* p. 93.

15. Merleau-Ponty, "Cezanne's Doubt," in *Essential Writings of Merleau-Ponty,* p.
244.

16. For the explicit distinction (though with a slightly different terminology), see
Heelan, "Experiment and Theory," p. 515. For a review of literature on hermeneu-
tics, see Josef Bleicher, *Contemporary Hermeneutics* (Boston: Routledge and Kegan
Paul, 1980). See also Heelan, *Space-Perception,* p. 194; "Being-in-the-World," and
"Natural Science as a Hermeneutics of Instrumentation," pp. 181–204.

17. Quoted in Mildred Portney Chase, *Just Being at the Piano* (Culver City, Calif.:
Peace Press, 1981), p. 37.

18. From "The Context of Performance," in *Actors on Acting,* ed. Toby Cole and
Helen K. Chinoy (New York: Crown, 1970), p. 668.

19. Mildred Chase, *Just Being at the Piano* (Culver City, Calif.: Peace Press, 1981),
pp. 24–25, 30.

20. Aristotle, *Nichomachean Ethics,* VI, chapt. 5, 1140b, trans. William D. Ross, in
The Basic Works of Aristotle, ed. Richard McKeon (New York: Random House, 1941),
pp. 1026–27.

21. Note that I am not discussing *phronesis* here, as Caputo does, in the context
of paradigm shifts. The role that phronesis—both on the part of experimenters
and theorists, and the distinction, neglected by Caputo, is crucial—might have in
such shifts deserves further consideration. It is incorrect to say, with Caputo, that
moments of paradigm change are moments "of free play and intellectual legroom,"
for paradigm shifts are cases of shifting from one already established position to
another (Kuhn, *Essential Tension,* p. 77). By then, the legroom has already been
explored. See Caputo, *Radical Hermeneutics,* p. 222.

22. Crease and Mann, *Second Creation,* pp. 210–14.

23. For a brief popular account of the science in cigarette lighters, see Robert P. Crease, "From a Spark to a Flame," *Atlantic* 258 (September 1986): 94–95.

24. Leon M. Lederman, "High Energy Experiments," in *Gauge Theories in High Energy Physics*, eds. Mary K. Gaillard and Raymond Stora (Amsterdam: North-Holland, 1983): 829–63.

25. The incident is recounted in Robert P. Crease, "Mea Culpa in the Lab," *Columbia* 12 (February 1987): 41.

26. Gadamer, *Truth and Method*, p. 99.

27. Ibid.

28. Hickman develops Dewey's position on this issue in *Dewey's Pragmatic Technology*, chapter 5.

29. Crease, "History of Brookhaven National Laboratory, Part One," p. 185.

30. Robert P. Crease, "The History of Brookhaven National Laboratory, Part Two: The Haworth Years," *Long Island Historical Journal* 4 (1992): 3–48.

31. Quoted in Crease and Mann, *Second Creation*, pp. 388–89.

5. Performance: Representation

1. For an exhaustive introduction to traditional approaches to theory, see Frederick Suppe, "The Search for Philosophic Understanding of Scientific Theories," in *The Structure of Scientific Theories*, ed. Frederick Suppe (Chicago: University of Illinois Press, 1974): 3–241; for a bibliography on theories, compiled by Rew E. Godow, Jr., see the same work, pp. 615–46. See also Percy Bridgman, *The Logic of Modern Physics* (New York: Macmillan, 1927); Carl G. Hempel and Paul Oppenheim, "Studies in the Logic of Explanation," *Philosophy of Science* 15 (1948): 135–75; Norwood Hanson, *Patterns of Discovery* (Cambridge: Cambridge University Press, 1958); Arthur Danto and Sidney Morgenbesser, *Philosophy of Science, Readings* (New York: Meridian, 1960); Nagel, *Structure of Science*, as well as various volumes of the series Minnesota Studies in the Philosophy of Science.

2. Some theatre consists of freely improvised performances in which actors, without much apparent forethought, spontaneously and individually apply skills to whatever features of the environment inspire them—remarks and gestures of passersby, noises, and the like. Here, each performance naturally differs from all others. In most theatre, however, the content of performances is codified to some extent in the form of a script, and even "pure" improvisation has its constraints, and usually consists of blocks of previously formed material loosely connected by short improvisational sections.

3. Heelan, "After Experiment." Heelan points to Hilbert's well-known lecture, "*Axiomatisches Denken*," in David Hilbert, *Hilbertiana: Fünf Aufsätze* (Darmstadt: Wissenschaftliche Buchgesellshaft, 1964): 1–11, as the classical expression of this view.

4. Wilshire, *Role-Playing and Identity*, p. 86.

5. It is unlikely, for instance, that a graduate student in physics would recognize these equations in the form in which Maxwell originally wrote them down. For an example of an historical treatment of Maxwell's equations, see Chien Ning Yang, "Maxwell's Equations, Vector Potential and Connections on Fiber Bundles," in *Proceedings of The Gibbs Symposium*, ed. George D. Mostow and D. G. Caldi (Providence, R.I.: American Mathematical Society, 1990), pp. 253–54.

6. Crease and Mann, *Second Creation*, chapt. 7.

7. See, for instance, Roger Sessions's discussion of this point in *Questions About Music* (New York: W. W. Norton, 1970), chapt. 3. Let disbelievers try to follow their folio editions of Shakespeare next time they watch one of that author's plays in performance, or follow an original operatic score in a staging of the work.

8. Thomas Huxley, in *Biogenesis and Abiogenesis,"* in *Collected Essays* (New York: Olms, 1970), p. 244.

9. Crease and Mann, *Second Creation*, pp. 386–89.

10. Wilshire, *Role-Playing and Identity*, p. 89.

11. In Eugene Wigner, *Symmetries and Reflections* (Woodbridge, Conn.: Ox Bow Press, 1979), p. 237. See also Mark Steiner, "The Application of Mathematics to Natural Science," in *The Journal of Philosophy* 86 (1989): 49–480, which cites some similar expressions of bafflement concerning the privileged role of mathematics in natural science, but which makes no more headway than Wigner in elucidating it.

12. James R. Newman, *The World of Mathematics*, vol. 3 (New York: Simon and Schuster, 1956), p. 1534.

13. Felix Klein, "Vergleichende Betrachtungen über neuere geometrische Forschungen" (Erlangen, 1872), reprinted in *Mathematische Annalen* 43 (1893): 63–100.

14. For a narrative account of the growing role of symmetries and invariances in modern physics, see Crease and Mann, *Second Creation*, chapts. 10-15.

15. For an account of this episode, see ibid., chapt. 14.

16. Quoted in ibid., p. 187.

17. Ibid., p. 241.

18. See Robert P. Crease and Charles C. Mann, "Yogi and the Quantum," 302–14 (from which much of this discussion of quantum mysticism is taken); also Sal Restivo, *The Social Relations of Physics, Mysticism, and Mathematics* (Dordrecht: Reidel, 1985).

19. Erwin Schrödinger, "The Current Situation in Quantum Mechanics," *Die Naturwissenschaften* 23 (1935): 812.

20. Niels Bohr, *Atomic Theory and the Description of Nature* (Cambridge: Cambridge University Press, 1934), p. 54.

21. For an introduction to the work of philosophers of science on quantum puzzles, see James T. Cushing and Ernan McMullin, eds., *Philosophical Consequences of Quantum Theory* (Notre Dame: University of Notre Dame Press, 1989).

22. Fritjof Capra, *The Tao of Physics: An Exploration of the Parallels Between Modern Physics and Eastern Mysticism.* (Berkeley: Shambhala Publications, 1976), p. 298; Gary Zukav, *The Dancing Wu Li Masters* (New York: Morrow, 1979). Among the numerous other authors who espouse quantum mysticism is Alex Comfort (of *Joy of Sex* fame), in *Reality & Empathy: Physics, Mind & Science in the 21st Century* (Buffalo: State University of New York Press, 1984). Quantum mysticism has even appeared in movies; the baseball groupie played by Susan Sarandon in *Bull Durham* is a quantum mystic.

23. Zukav, *Wu Li Masters*, p. 28.

24. Shirley MacLaine, *Dancing in the Light* (New York: Bantam, 1985).

25. See Crease and Mann, "Yogi and the Quantum."

26. Restivo, *Social Relations of Physics*, p. 22.

27. Richard Feynman, *The Character of Physical Law* (Cambridge, Mass.: MIT Press, 1965), pp. 39–40.

28. Zukav, *Wu Li Masters*, p. 31; Capra, *Tao of Physics*, p. 5.

29. Patrick A. Heelan, "The Quantum Theory and the Phenomenology of Social-Historical Phenomena," unpublished manuscript.

30. Ibid. For other discussions of this path-dependent approach to quantum phenomena, see Heelan, "Complementarity, Context Dependence, and Quantum Logic," in *Foundations of Physics* 1 (1970): 95–110.

31. Stephen Jay Gould, *Wonderful Life: The Burgess Shale and the Nature of History* (New York: W. W. Norton, 1989), p. 35.

32. Ibid., p. 48.
33. Ibid., p. 51.
34. Ibid., p. 51.
35. As many anecdotes attest: "If you're an experimenter, you get the illusion that the theorists are all such *smart* bastards," University of London physicist Dick Learner once remarked. "But many theorists have no idea what's going on in an experiment. If you stand a theorist next to an apparatus, it breaks . . . [and] a few [are] really banana-fingered gentry. . . . The point to all this is that you [i.e., the experimenter] must remember that you know more about what you are seeing and how you are seeing it than they do. Experimental physics, alas, has an inferiority complex." Quoted in Crease and Mann, *Second Creation,* p. 119.
36. "For every good performer, the role of the score undergoes a transformation when it ceases to be a theory and becomes instead a mnemonic, then the artist's scorebook becomes a set of 'places' or topoi, the function of which is to remind the artist of the suites and sequences to be performed. As such it is a local, personal, contextual, historical, technological, and artistic guide, it is an open or endless set of memory cues, it is no longer a universal theoretical prescription." (Heelan, "Experiment and Theory," p. 522). The codification, to be performed, depends on accepted, standardized laboratory techniques; hence the "praxis-ladenness" of theory. How theoretical scripting works out in the case of the human sciences demands a more elaborate treatment than is possible here.
37. Frank Wilczek, "What Did Bohr Do?" in *Science* 255 (1992): 346.
38. Stephen G. Brush, *The History of Modern Science* (Ames: Iowa State University Press, 1988), p. 409.
39. For more on quantum mechanics, see Heelan, "Quantum Theory."

6. Performance: Recognition

1. Thomas Kuhn, for instance, exposes many of these limitations in chapter 6 of *The Structure of Scientific Revolutions* (Chicago: University of Chicago Press, 1962). Kuhn remarks, for instance, that "Though undoubtedly correct, the sentence, 'Oxygen was discovered,' misleads by suggesting that discovering something is a single simple act assimilable to our usual (and also questionable) concept of seeing. That is why we so readily assume that discovering, like seeing or touching, should be unequivocally attributable to an individual and to a moment in time" (Ibid., p. 55).
2. See for instance Otto Glasser, *Wilhelm Conrad Roentgen and the Early History of the Roentgen Rays* (Baltimore: Charles C. Thomas, 1934).
3. Ibid., chapt. 2.
4. "X Ray," in *Encyclopaedia Britannica* (Chicago: Encyclopaedia Britannica, 1992), p. 791.
5. Glasser, *Wilhelm Conrad Roentgen,* pp. 222–23.
6. As, for instance, Arthur Koestler did in *The Act of Creation* (New York: Macmillan, 1964).
7. N. R. Hanson, "The Logic of Discovery," in *Science: Men, Methods, Goals,* eds. Baruch Brody and Nicholas Capaldi, (New York: W. A. Benjamin, 1968), pp. 150–62, and Peter Achenstein, "Inference to Scientific Laws," in *Minnesota Studies in the Philosophy of Science,* vol. 5 (Minneapolis: University of Minnesota Press, 1970), pp. 87–111.
8. Auguste Comte, *The Positive Philosophy of Auguste Comte,* trans. Harriet Martineau (New York: Blanchard, 1858), p. 38.
9. Rudolf Carnap, "Logical Foundations of the Unity of Science," in *Foundations of the Unity of Science* vol. 1, eds. Otto Neurath, Rudolf Carnap, and Charles Morris (Chicago: University of Chicago Press, 1971), p. 49.

10. Plato, *Meno*, 80d, trans. by William K. C. Guthrie, in *The Collected Dialogues of Plato*, eds. Edith Hamilton and Huntington Cairns (Princeton: Princeton University Press, 1971), p. 363.

11. For recognition of objects and events, see for instance 1452a35–7; in the *Prior Analytics*, Aristotle refers to recognition of rules.

12. Edward Casey, *Remembering* (Bloomington: Indiana University Press, 1987): 122–41; H. H. Price, *Thinking and Experience* (Cambridge: Harvard University Press, 1953), chapt. 2; Kenneth M. Sayre, *Recognition: A Study in the Philosophy of Artificial Intelligence* (Notre Dame: University of Notre Dame Press, 1965); J. O. Urmson, "Recognition," *Proceedings of the Aristotelian Society* 61 (1956): 259–80.

13. On Hegel and recognition, see Elliot L. Jurist, "Hegel's Concept of Recognition," *The Owl of Minerva* 19 (1987): 5–22.

14. On the Eightfold Way, see Crease and Mann, *Second Creation*, chapt. 14.

15. Robert K. Merton, "The Unanticipated Consequences of Purposive Social Action," *American Sociological Review* 6 (1936): 894–904. See also Robert P. Crease and Nicholas P. Samios, "Managing the Unmanageable," *Atlantic* 267 (January 1991): 80–88.

16. See Merleau-Ponty, *Phenomenology of Perception*, p. 185.

17. John Dewey, *Art as Experience* (New York: G. P. Putnam's Sons, 1958), p. 52.

18. For a popular review of some cases of serendipity and some of the literature, see George B. Kauffman, "Oops . . . Eureka!", *1991 Yearbook of Science and the Future* (Chicago: Encyclopaedia Britannica, Inc, 1990): 225–40.

19. Casey, *Remembering*, pp. 130–32.

20. Arthur Koestler, *Insight and Outlook* (London: Macmillan, 1949), chapt. 18.

21. George K. T. Conn and Henry D. Turner, *The Evolution of the Nuclear Atom* (New York: American Elsevier, 1965), p. 135.

22. John L. Heilbron, "The Scattering of α and β Particles and Rutherford's Atom," *Archive for History of Exact Sciences* 4 (1967): 247–307.

23. Casey, *Remembering*, pp. 123–24.

24. Leon D. Harmon, "The Recognition of Faces," *Scientific American* 229 (November 1973): 70–82.

25. For pathological science and its symptomology, see Charles Babbage, *Reflections on the Decline of Science in England, and on Some of Its Causes* (London: B. Fellowes, 1830), chapt. 5; Irving Langmuir, "Pathological Science," trans. Robert N. Hall, *Physics Today* 42 (October 1989): 36–48; Robert P. Crease and Nicholas P. Samios, "Cold Fusion Confusion," *New York Times Magazine*, September 24, 1989, pp. 34–38.

26. The scientist, Bernhard Meier, is quoted in *The New York Times*, March 21, 1991, sect. A, p. 10.

27. The scientific paper that resulted is by Bernhard Meier and Roland Albignac, in *Folia Primatologica* 56 (1991): 57–63.

28. Watson, *The Double Helix*, Norton Critical Edition, ed. Gunther Stent (New York: Norton, 1980).

29. Sigmund Freud, "Recommendations for Physicians on the Psycho-Analytic Method of Treatment," in *Collected Papers* (New York: Basic Books, 1959) 2:324.

30. Ernst Mayr, *The Growth of Biological Thought* (Cambridge: Harvard University Press, 1982), pp. 30–32.

31. For instance, see "Physics Parts the Red Sea" (no author), in *Science* 256 (1992): 246.

7. Performance and Production

1. See, for instance, Robert P. Crease, "Images of Conflict: MEG vs. EEG," *Science* 253 (1991): 374–75.

2. My attention was drawn to the importance of this concept by Zev Trachtenberg, "A Theory of Drama," (M.Phil. thesis, University College, London, 1980).

3. See Crease, "History of Brookhaven National Laboratory, Part One."

4. As is evident, for instance, from the news and research news stories about this process in *Science:* Robert R. Crease, "Choosing Detectors for the SSC," *Science* 250 (1990): 1648–50; David P. Hamilton, "Showdown at the Waxahachie Corral," *Science* 252 (1991): 908–1010; David P. Hamilton, "Ad Hoc Team Revives SSC Competition," *Science* 252 (1991): 1610; David P. Hamilton, "A New Round of Backbiting over the Cancellation of L*," *Science* 252 (1991): 1775.

5. "Impact of Large-Scale Science on the United States," *Science* 134 (1961): 161–64. Yale historian Derek de Solla Price adopted the phrase in a 1962 lecture series at Brookhaven National Laboratory, "Little Science, Big Science," subsequently published as a book, *Little Science, Big Science* (New York: Columbia University Press, 1963).

6. Alvin Weinberg, *Reflections on Big Science* (Cambridge: MIT Press, 1967), p. 67.

7. Weinberg had before him only two models of Big Science, large particle accelerators and the manned space program, neither of which had really matured. Forefront particle accelerators could still be built at universities and Project Mercury was in its infancy; Weinberg's article was based on an address given before a meeting of the American Rocket Society in Gatlinburg, Tenn., on May 4, 1961, the day before Alan Shepherd became the first American astronaut to be launched into space.

8. The example is from Michael Jochim, *Strategies for Survival* (New York: Academic Press, 1981), p. 11. I am indebted to Marshall Spector for drawing my attention to this reference.

9. Alvin Weinberg, "The Axiology of Science," *American Scientist* 58 (November-December 1990): 612–17.

10. *Large Nondefense R and D Projects in the Budget: 1980–1996* by David Moore and Philip Webre (Washington, D.C.: U.S. Congress, U.S. House of Representatives, Congressional Budget Office, July 1991).

11. This, for instance, is the point made by John A. Remington in "Beyond Big Science in America: The Binding of Inquiry," *Social Studies of Science* 188 (1988): 45–72, though he uses Weinberg's distinction between internal and external criteria.

12. For a discussion of an example, concerning the decline of administrative contracts in national laboratories, see Robert P. Crease and Nicholas P. Samios, "Managing the Unmanageable," *Atlantic* 267 (January 1991): 80–88.

13. See, for instance, H. Schmied, "A Study of Economic Utility Resulting from CERN Contracts" (Geneva: CERN 75-5, 1975; Second Study, CERN 84-14, 1984); Edwin Mansfield, "Estimates of the Social Returns from Research and Development" (paper presented at the Science and Technology Policy Colloquium of the American Association for the Advancement of Science, Washington, D.C., April 12, 1991).

14. See Lawrence Stone, "The Revival of Narrative," in *Past and Present* 85 (November 1979): 3–24.

15. See ibid.

16. Crease, "History of Brookhaven National Laboratory, Part One," pp. 167–88.

17. The use of fictionalized history in the philosophy of science is admitted with a certain chagrin by Herbert Feigl in "Beyond Peaceful Coexistence," in *Minnesota Studies in the Philosophy of Science,* vol. 5, ed. Roger Stuewer (Minneapolis: University of Minnesota Press, 1970), p. 3. But John J. C. Smart defends the use of fictionalized history in "Science, History, and Methodology," *British Journal for the Philosophy of Science* 23 (1972): 266; "Methodologists need examples from the history of science only because it is too hard to think up fictitious ones. It does not matter, therefore, whether the history is quite true."

Conclusion

1. See Robert P. Crease, "Top Scientists Must Fight Astrology—Or All Of Us Will Face The Consequences," *The Scientist* 3 (March 6, 1989): 9, 11.

2. Weissman, *They All Laughed at Christopher Columbus* (New York: Times Books, 1987), p. 270.

3. Antonin Artaud, *The Theatre and its Double,* trans. Mary Caroline Richards (New York: Grove Press, 1958), p. 41.

4. Hans Reichenbach, *Philosophic Foundations of Quantum Mechanics* (Berkeley: University of California Press, 1982), p. vi.

INDEX

Adorno, Theodor: 69
aesthetics: 14, 45, 108–109, 120, 181
analogy: theatrical, 4, 6, 13, 15, 29, 33, 74, 77, 95–102, 122, 158, 164–67, 171–72, 177, 178–79, 184; use of, 5–6, 37, 74–77; between scientific and ordinary perception, 77–95, 145, 152; between narrative and performance, 174–77
anti-empiricism: 28, 89–91
aporia: 66, 150
Aristotle: on first principles, 3–4; Aristotelian mechanics, 18–19; on philosophical activity, 22; on *techne*, 40; on metaphor, 75; on substance, 86; on *phronesis*, 114–15; on recognition, 147–51
Artaud, Antoine: 182
artistry of experiment: 12, 14, 107–20

Bach, C.P.E.: 114
Bacon, Francis: 96
Bernstein, Jeremy: 5, 77
"Big Science": 167–72
Black, Max: 75
Blackett, P.M.S.: 80
Blewett, John: 120
Bohr, Niels: 133
book of nature: 19, 24, 53, 139, 184
Broad, William: 7, 8
Bronk, Detlev W.: 120
Brookhaven National Laboratory: 120, 162
Brush, Stephen: 140

Capra, Fritjof: 134–36
Caputo, John: 98
Carnap, Rudolf: 16, 28, 89, 146
Cartan, Élie-Joseph: 128
Casey, Edward: 148, 152–53
Chaikin, Joseph: 114
Chase, Mildred: 114
Cohen, Morris R.: 3
Comte, Auguste: 146
context of discovery/justification: 94, 117
Copenhagen interpretation: 133, 139–40
Crookes, Sir William: 143
cultural attitude of science: 99, 150, 156, 157, 158
Curie family: 74, 116, 117, 149

data: foundational, 3, 18, 28; and phenomena, 86, 90–91; as constraining theory, 90;

do not exhaust experimentation, 101; path-dependency of in quantum phenomena, 136–37; denominate presence of phenomena, 86, 101, 159; fluidity of, 91, 159, 167
Demiurgic knowledge: 13, 19, 140, 181
Descartes, René: 25
Dewey, John: 14, 23–24, 36–46, 51, 58, 61, 68, 72, 101, 103, 106, 120, 134, 150, 151, 179
Dirac, P.A.M.: 92, 129
discovery: logic of, 14, 145; knowledge a discovery, in spectator's view, 23, 37; does not pop out of equipment, 36, 119; Dewey on, 42, 45; artistry of, 70–71; as recognition, 83, 142–57; missed, 111, 116–18; pre-history of a, 143, 146
dual semantics of science: 87, 89, 130–31

Einstein, Albert: 7, 8, 20, 21, 23, 30, 55, 132
empiricism: 28, 47, 90–91
epiphany of nature: 184
epistemology: 1, 2, 14, 133, 140, 141
experiment: neglect of, 2–6; value of an inquiry into, 6–15, 180–85; failed, 7; antinomic character of, 9, 11, 31, 94, 98, 164–65, 179; mythic account of, 16–22, 23, 26, 28, 29, 37, 41, 44, 54, 67, 68, 78, 95, 96, 108, 117, 134, 155, 176, 180, 181, 182, 185; philosophers and, 22–29, 184–85; philosophical tools needed to study, 29–33, 36–73; theatrical analogy with, 4–6, 95–102, 178–79; technology and artistry of, 70, 107–21; theory and, 122–41; discovery and, 142–57; production of, 160–66; narratives and, 172–77

Fermi, Enrico: 109, 116, 138
Feynman, Richard: 1, 20–21, 39, 135
Fleck, Ludwik: 100, 109, 183
Forman, Paul: 8
Franklin, Alan: 4, 21
fraud: 7–8, 12, 154
Frege, Gottlob: 109, 120
Freud, Sigmund: 5, 34, 35, 156

Gadamer, Hans-Georg: 57, 69, 118–19
Galileo: 2, 184; alleged Leaning Tower experiment, 18–19; discovery of isochrony of pendulum, 31–32, 65; Husserl on, 52–57, 71–72
Galison, Peter: 4

ROBERT P. CREASE is Assistant Professor of Philosophy at the State University of New York at Stony Brook and coauthor (with Charles C. Mann) of *The Second Creation: Makers of the Revolution in Twentieth-Century Physics.*